Destination London: The Expansion of the Visitor Economy

Edited by
Andrew Smith and Anne Graham

University of Westminster Press
www.uwestminsterpress.co.uk

Published by
University of Westminster Press
115 New Cavendish Street
London W1W 6UW
www.uwestminsterpress.co.uk

First published 2019

Cover and back cover image: Diana Jarvis; Front cover image: Tristan Luker
Printed and bound by CPI Group (UK) Ltd, Croydon, CR0 4YY
Print and digital versions typeset by Siliconchips Services Ltd.

ISBN (Paperback): 978-1-912656-26-4
ISBN (PDF): 978-1-912656-27-1
ISBN (EPUB): 978-1-912656-28-8
ISBN (Kindle): 978-1-912656-29-5

DOI: https://doi.org/10.16997/book35

The full text of this book has been peer-reviewed to ensure high academic standards. For full review policies, see: http://www.uwestminsterpress.co.uk/site/publish.

Suggested citation:
Smith, A. and Graham, A. (eds.) 2019. *Destination London: The Expansion of the Visitor Economy.* London: University of Westminster Press. DOI: https://doi.org/10.16997/book35 License: CC-BY-NC-ND 4.0

To read the free, open access version of this book online, visit https://doi.org/10.16997/book35 or scan this QR code with your mobile device:

Contents

Dedication

This book is dedicated to Simon Curtis, who died suddenly in January 2019 just after the final manuscript was submitted to the publishers. Simon was a much-loved and much-admired member of the Tourism and Events team at the University of Westminster and he will be greatly missed by everyone who worked with him. He was an enthusiastic champion of the project to write a book about tourism in London and we are so sad that he never got to see the text published. Simon joined the University in 2011 after a distinguished career in tourism consultancy and destination management, working as Head of Tourism at Kent County Council and Head of Tourism and Heritage at Medway Council. He was a passionate advocate for heritage and the restoration of historic buildings. Simon passed this enthusiasm on to the next generation through his Heritage Tourism module, which was extremely popular with our students. The chapter on the River Thames that Simon wrote in this volume is typical of his work in its thoughtfulness, its clarity and its obvious passion for the subject matter. Since his death, the first lines of Simon's chapter – taken from a Song by Wallace and O'Hogan – have taken on a new resonance:

'Kingdoms may come, kingdoms may go; whatever the end may be, Old Father Thames keeps rolling along; down to the mighty sea'.

Simon, this book is dedicated to you.

Acknowledgements

This book has been produced collaboratively by members of the Tourism and Events Research Group at the University of Westminster and we would like to thank members of this Group – past and present – for their contributions. Tourism research at Westminster has a significant pedigree. For several decades the University has hosted a group of tourism researchers within a School dedicated to the study of architecture and related built environment disciplines (transport, planning, construction, urban design). Analysing tourism from an urban planning and/or place-based perspective is something unique to Westminster and this approach underpins the work presented here. We would also like to thank Jayni Gudka for her insights into social tourism in general, and Unseen Tours in particular, that have informed Chapter 7.

The idea for an edited collection on Destination London was first mooted back in 2009, so it has taken a decade to come to fruition. Several people have helped to make this idea a reality. Andrew Lockett at the University of Westminster Press has provided unwavering support since we first suggested the idea of a book on tourism in London. We would also like to thank the external reviewers who provided valuable feedback at various stages. Tristan Luker gave us permission to use some of his brilliant images of London to illustrate the book, including the one on the front cover. Thank you, Tristan – you can see more of his work at www.tristanluker.com. Two other people also contributed images for the book. Mason Edwards (www.masonedwardsdesign.co.uk)

produced some of the figures and diagrams, and one of our former students, Eman Mustafa, kindly agreed to let us use several of her photographs. Thanks you to both. We would also like to thank the School of Architecture + Cities for funding publication costs including the costs of several images licensed from the RIBA Collection.

The book's editors and contributors are pleased to declare that they have no competing interests though, as noted in the biographical profiles at the end of the book, some have undertaken consultancy work either as individuals or Westminster staff members. Jayni Gudka's contribution to Chapter 7 reflects her experience leading Unseen Tours' Responsible Tourism and Communications work.

CHAPTER 1

Destination London: An Expanding Visitor Economy

Andrew Smith

Introduction

No city in the world is better covered by literature – fictional and non-fictional – than London. From Pepys, via Dickens, to Ackroyd, London has benefited from a series of talented historians, novelists and commentators who have provided detailed accounts of the city's condition. In the past few years a new tranche of books has been published on the contemporary character of the UK capital: with Anna Minton's *Big Capital*, Rowan Moore's *Slow Burn City*, Ben Judah's *This is London* and Iain Sinclair's *The Last London* notable examples. One thing that unites these otherwise excellent accounts is the conspicuous absence of discussions about the city's visitor economy. This is a notable omission, given the scale and significance of tourism in London. Over the years, the city has earned various nicknames that purport to represent its essential nature: 'the great wen'; 'the big smoke'; 'the city of villages'. But the epithet that perhaps best represents contemporary London might be: 'the city of tourists' or *Destination London.*

London hosts a very significant visitor economy and overnight visitors contribute approximately £14.9 billion of expenditure to the city every year

---

How to cite this book chapter:

Smith, A. 2019. Destination London: An Expanding Visitor Economy. In: Smith, A. and Graham, A. (eds.) *Destination London: The Expansion of the Visitor Economy.* Pp. 1–13. London: University of Westminster Press. DOI: https://doi.org/10.16997/book35.a. License: CC-BY-NC-ND

(London and Partners 2017). When the city hosted the Olympic and Para-lympic Games in 2012 the UK's capital was already a leading global destination, but staging this mega-event instigated a new period of growth. In the period 2011–2016 tourism numbers increased by 25 per cent and over the past few years the city has experienced a series of record years for arrivals. Despite already being one of the three most visited cities in the world – hosting 31.2 million overnight visitors in 2016 – city officials expect visitor arrivals to increase further: to over 40 million overnight visitors by 2025 (London and Partners 2017). Put simply, tourism is already a very significant economic and social phenomenon in London, but over the next few decades it will become even more pivotal and pervasive.

London's status as one of the world's most visited destinations is not universally welcomed. At the moment there is considerable media and academic attention dedicated to the problem of rapid tourism growth and what has become known as *overtourism*. This coverage has focused on various European capitals: from Berlin to Barcelona, Ljubljana to Lisbon. Even though the UK's capital city seems like the ideal case through which to explore the ways that destinations evolve and expand, there has been surprisingly little attention devoted to London in these debates. This book explores how and why tourism is growing in Europe's most popular city destination; and what benefits and problems accrue from expanding the tourism sector in a city already hosting 19 million overseas tourists and 12 million overnight domestic visitors every year. These additional people mean London's population grows considerably every day, especially when one considers the 300,000 people that commute daily to the capital from outside Greater London and the daily influx of 750,000 non-staying visitors. London hosts a residential population of around 8.8 million people, but its 'daytime' population, i.e. that which includes workers, visitors and tourists, is estimated to be over well over 10 million (GLA 2015). Put another way, tourists and day visitors now make up over 10 per cent of the people who inhabit London every day.

The book analyses how and *why* the expansion of the visitor economy is happening and what effect this is having on the city. Contributions from various authors demonstrate how Destination London is developing through the extension of tourism into new spaces and new spheres. The book outlines how parts of London not previously regarded as tourism territories, e.g. residential suburbs, peripheral parks and private homes, are now subjected to the tourist gaze. Tourists are being encouraged to visit places outside the centre and stay in accommodation owned by residents. In a similar manner, London is constantly creating new eventscapes to capitalise on the experience economy and providing reasons to visit at different times – in winter and at night. These types of initiatives feature prominently in London's new 'Tourism Vision', which explicitly outlines the city's aim to grow tourism 'by encouraging visitors to explore the city's outer districts, both in and out of season and around the clock' (London and Partners 2017, 16). This ambition is nothing new. A guidebook published

Figure 1.1: Tourism is Expanding beyond London's Tourist Bubble (Photo: Tristan Luker).

in 1978 lamented that many tourists miss out on experiencing London's 'multiple fascination', because they failed to go beyond the West End and conventional tourist attractions ... 'if only they moved to the right or the left of those well beaten tracks' (Crookston 1978, 8).

Contemporary expansion is being facilitated by extending the capacity of existing services (e.g. by running the Underground 24 hours a day), and by building new infrastructure (e.g. the new Crossrail network and a new runway at Heathrow Airport) and accommodation provision (plans for 23,000 new hotel rooms by 2025). However, growth in the visitor economy is driven more by market and cultural trends than any deliberate planning and policy; and this unfettered growth is likely to outrun formal provision. The rise of social media and the sharing economy, and the desire for new, distinctive and personalised experiences, are pushing tourists into peripheral locations, but also advancing tourism into spheres not normally considered tourism territory. Growth is likely to be enabled and absorbed by unofficial tourism providers, including London's residents who now provide a range of services: most obviously accommodation, but also food, travel and guiding. This book explores these trends and, in doing so, highlights the mechanisms and processes that are driving the expansion of the visitor economy. The discussion enhances the understanding of London, but it also helps us to better appreciate the ways that tourism in cities is expanding into new spaces, times and spheres.

Tourism Expansions and Extensions

How cities grow and develop is an established field of academic enquiry. A large number of texts explain how cities change over time, including detailed consideration of the processes of regeneration and gentrification. London is a city known for its planned growth – its expansion has been carefully orchestrated, and the limits of the city are still bounded by a 'green belt' (See Figure 1.2). But urban growth and development also happens in unplanned and/or unofficial ways: planning policy is breached through various types of informal, illegal and unsanctioned development. In the twenty-first century, attempts to curtail urban sprawl mean expansion is often vertical, rather than merely horizontal, a trend explored in Chapter 6 of this text. Alongside analysing the expansion of cities, it is important to analyse how existing urban areas evolve and Tim Butler's (1997) work on the waves of gentrification in London is very important in this regard. Various forces driving urban change can be understood as cycles of development involving cultural pioneers paving the way for more mainstream, mass market clientele. This model is equally relevant to urban change instigated by increased demand for tourism services. The visitor economy is now acknowledged as a force that shapes cities, but it is rarely analysed in depth as a significant contributor to urban transformations. Dedicated analyses which explain how tourism develops and expands in cities are even rarer. In the era of 'the entrepreneurial city', the visitor economy has become central to the economy and life of many cities and it deserves more consideration.

This book is situated within an emerging body of work that appreciates the way tourism has diversified, making it harder to separate from other activities.

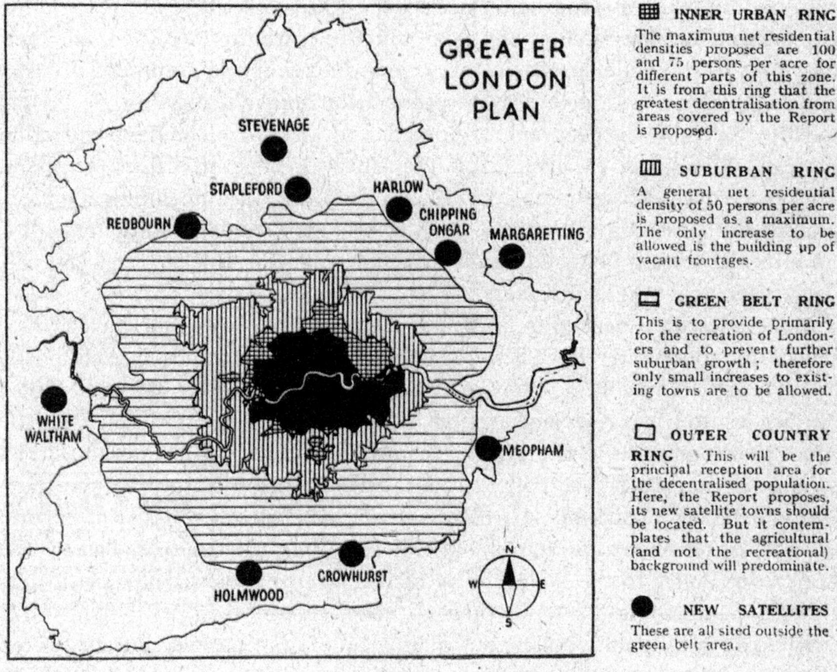

Figure 1.2: Greater London Plan's Four Zonal 'Rings' (Abercrombie/RIBA Collections, 1945).

Tourism *in cities* exhibits particularly pronounced indivisibility (and therefore, invisibility) due to the way that tourism activities and other forms of consumption and mobility coincide. In a city like London it is very hard to distinguish tourists from other mobile elites, including students, those travelling for business and people that reside in multiple locations. The traditional distinction between tourists and locals is increasingly blurred, something exascerbated by noted role reversals. In contemporary cities – particularly global tourism cities – it seems as though tourists want to be locals and locals want to be tourists (Lim and Bouchon 2017). For this reason, it is helpful to talk about tourism as a role or set of behaviours that people perform, rather than as merely a category of consumer defined by where someone lives. This allows for consideration of 'as if' tourists – residents who use tourist services/spaces and act like tourists even if they live locally (Novy 2018), alongside tourists who use residential services/ spaces and behave like residents (Maitland 2009).

Traditionally, tourism has been regarded as something that takes place in a distinctive part of a city – the 'tourist bubble' or 'entertainment district' – but it increasingly infiltrates a more diverse set of urban spaces and places. This process of 'tourism territorialisation' is analysed in this book. The growing literature on city tourism often equates this with the process of touristification – 'the coming into being of a touristic place' (Stock 2007, 3) – where the city and the citizenry are appropriated as objects of tourism consumption. There are obvious links to the wider processes of commodification and commercialisation and to the aforementioned ideas of regeneration and gentrification. According to Novy (2018, 7) 'the geographical spread of tourism in Berlin has occurred in parallel with a spatial expansion and intensification of gentrification processes'. This pattern is also noticeable in other European cities, including London. Non-central areas – for example, the South Bank, Camden and Spitalfields – have fundamentally changed because of their appeal to international tourists, day visitors and other consumers.

More research is needed to understand how urban/residential space is converted into tourism territory. This is a complex process; and it is over-simplistic to suggest tourism commodifies, or commercialises space. As Biddulph (2017, 32) argues, tourism does not expand into empty or moribund space: as 'the space that tourism is territorialising from the centre out is already the site of a range of commercial activities'. This radial expansion of tourists and tourism is explored in this book, which examines the spread of tourism beyond established tourism zones into peripheral, suburban and residential areas. The book contributes to the literature by focusing on the ways that tourism territory expands in, and extends through, the contemporary city. Expansion is partly driven by public policy, but mainly by tourist markets and associated capital which are constantly seeking new 'products' to exploit. Understanding this expansion helps us to understand the ways in which urban areas are commodified and consumed.

The Development of London Tourism

Tourism in London is a very established activity/industry with a long history. There is insufficient space here to provide a detailed historical account, but a brief review of the emergence of tourism in the modern era provides a helpful introduction to the chapters that follow. In this endeavour we are grateful for the work of David Gilbert who has produced a range of articles on the 'under-acknowledged' role of tourism in the development of modern London (Gilbert 1999, 279). The history of tourism in London is significant for various reasons, not least because the dominant images that are shared and circulated of the city today are still heavily reliant on key periods in the past – particularly the Victorian era and the 'Swinging [Nineteen] Sixties'.

Whilst London's appeal is based on historical attractions that date back to Roman times, the city's tourism 'industry' arguably dates back to the nineteenth century. In the period 1820–1840 new facilities were established that still provide the backbone of the city's tourism sector: iconic attractions (London Zoo, Madame Tussauds), leisure settings (public parks, shopping streets) and supporting amenities (hotels and rail stations). Key institutions introduced at this time included the Royal Polytechnic Institution at 309 Regent Street (est. 1838) – now the University of Westminster – which was one of London's pioneering visitor attractions. This was a precursor to the city's Science Museum (est. 1857) – one that allowed new technologies and inventions to be shown to the public.

The Great Exhibition of 1851 is said to mark the birth of the modern tourism 'industry' not only in London, but globally. The pioneering entrepreneur Thomas Cook organised tours which brought thousands of people via train into the UK capital. This pivotal event – which like the Polytechnic Institution aimed to reassure people about modernity – highlights one of the factors underpinning London's enduring appeal: the way the city allows visitors a glimpse into the future. The city's history and heritage has always been a draw for visitors, but from the early nineteenth-century onwards tourists also visited London 'to see a new world in the making' (Gilbert 1999, 281). The appeal of London as a destination was never driven by its beauty or aesthetic qualities – it was (and still is) compared unfavourably to Paris, Berlin or Brussels in this regard (Gilbert 1999). What fascinated tourists was the entrepreneurial dynamism driven by its role as the centre of a trading empire. For example, Burton's (1996) account of the experiences of Indian visitors highlights that they were entranced by London's 'vitality and ceaseless motion'. Similarly, Gilbert's (1999) analysis of nineteenth century guidebooks suggests these texts were less concerned with architectural merit and more concerned with detailing the sheer pace of commercial life. The appeal of slumming – something discussed in detail by Claudia Dolezal and Jayni Gudka in Chapter 7 of the book – also meant that some voyeuristic tourists ventured into the impoverished easterly districts of Whitechapel and Stoke Newington. Therefore, whilst the appeal of peripheral neighbourhoods and everyday activities has certainly intensified in recent times, these also attracted London's tourists in the nineteenth century.

In the first half of the twentieth century, tourism in London was severely affected by war and various political and economic crises. The main drivers of tourism at this time included various international exhibitions such as the Franco–British Exhibition (1908), the Olympic Games (1908), the Japan – British Exhibition (1910) and the British Empire Exhibition (1924–25). Despite the damage inflicted by the Second World War, which ruined much of the city and public finances, London also staged the 1948 Olympic Games. Multiple benefits were sought from hosting this event but, according to one Foreign Office official, 'the government's main interest is to seize the occasion to develop the tourist trade' (Polley 2011, 132). The Games were used to generate publicity and demand for the newly created British European Airways, which was based at the recently opened London Airport at Heathrow. Other major events were also used to rejuvenate the city's morale and the built environment. Staging the 1951 Festival of Britain meant the derelict South Bank was transformed into a new cultural district featuring boldly designed galleries, theatres and concert halls.

The post-war period was a key time of global migration to London. As a liberal metropolis and a port city that functioned as the commercial and administrative fulcrum of a global Empire, London had long been a city of migrants. Economic migrants and persecuted groups from across Europe came to live in London, including Jewish and Huguenot refugees. But in the second half of the twentieth century London also experienced a significant influx of people from the Caribbean, Hong Kong, Pakistan, Bangladesh and India. These migrants added to the appeal of London as a world city, a place where a range of international foods, traditions and music could be experienced. Certain clusters, most notably Chinatown in central London, but also the Bangladeshi communities of inner east London, became attractive areas for visitors curious about London's cosmopolitanism. A guide book published in 1978 illustrates this dimension of the city's appeal:

> it is the immigrants to the capital who have given it such an exciting variety of ambience and taste… making Gerrard Street in Soho tinkle like a Kung Fu movie, Bayswater Cafés murmur with Austro–Hungarian melancholy, Islington delicatessens vibrate with vehement Italian (Crookston 1978, 8).

London's international credentials meant the city was visited not so much as the capital of England and the UK, but as a global capital. This was reinforced by key (colonial) attractions featuring world-wide collections, like the British Museum, Kew Gardens and London Zoo. This globalism provided the foundation for London's contemporary appeal, as evidenced by a recent tourism marketing campaign which featured the strapline 'See the World: Visit London'.

By the 1960s London had become a vibrant metropolis again, not only because of its enduring role as a centre for trade and commerce, but because of its role as the centre for popular culture. London began to adopt the modern urban aesthetics popularised by New York, perhaps best symbolised by the opening of the Hilton on Park Lane in 1963 (Czyzewska and Roper 2017). In this new age of liberation and leisure, London's appeal was based on its cutting edge 'scene' and 'buzz' – attributes inextricably associated with high profile musicians, artists, fashion designers, film directors and photographers. This era witnessed the rise of commodity culture with cultural producers and consumers concentrated in Soho, a creative cluster which positioned London within 'international networks of fashion promotion, popular culture, travel and tourism' (Gilbert 2006, 4). 'Swinging London' was dismissed as merely 'a few hundred exhibitionists with a flair for self-promotion' (Aitken 1967 cited in Gilbert 2006), but this representation was disseminated widely and proved to be remarkably enduring.

Despite the different layers of attraction that had emerged by the end of the 1960s – the frenzied commercial activity, the monumental edifices of state (and empire) and more latterly the cultural 'scene' – London tourism was still a relatively small-scale phenomenon compared to the multi-billion pound industry we see today. According to Tyler (2009, 418) 'tourism was never a particularly major part of London's economy until the 1970s when overseas arrivals doubled within a decade' – from 3 million to 6 million per annum. Boosted by a global appetite for international travel, and a series of urban renewal programmes focused on the city's docks and wharves (see Chapter 8), London began a 40-year period of (largely) uninterrupted tourism growth, which has endured to the present day (London and Partners, 2017). From the 1960s we also see the emergence of formalised tourism governance. The London Tourist Board was incorporated as a company in 1963 with an objective to manage tourism and promote London as a year-round tourist destination. In 1969 the company became one of 12 English tourist boards coordinated by the British Tourist Board.

At the end of the 1980s the first Tourism Strategy and Action Plan for London was developed by the Joint London Tourism Forum. However, the start of the next decade was a difficult period: the recessions that occurred at this time meant a decline of 1.5 million visits to London in the first two years of the 1990s (Church and Bull 2001). Subsequently, tourism arrivals grew very fast – buoyed by new forms of short break tourism, and the remarkable rise of Visiting Friends and Relatives (VFR) tourism which doubled 1991–1995 (Church and Bull 2001). This trend led to the fascinating realisation that, during the 1990s, 'people may have become the most important attraction, not the heritage and culture, but Londoners' (Church and Bull 2001, 148). The growth of VFR tourism has not only continued, it has accelerated with the latest figures suggesting VFR tourism makes up over 30 per cent of all overnight visits (London and Partners 2017). It is the growth of this market that has pushed tourists and tourism beyond the West End into more peripheral parts of London.

The twenty-first century has seen some significant changes, not least to the way tourism is managed and governed in London. Under the GLA Act (1999) the newly established elected Mayor of London was given the statutory responsibility for the promotion and development of tourism, which was then devolved to the London Development Agency (LDA). The LDA created a new organisation Visit London which effectively replaced the function of the London Tourist Board. These initiatives were instigated by the Labour Government 1997–2010, but when the Conservative–Liberal Democratic Coalition took over, they abolished the LDA and the other Regional Development Agencies. One result was that Visit London was folded into a new organisation – London and Partners – which also took on responsibility for other types of city marketing (to potential investors, students and film makers). This new organisational structure aimed to produce a more coherent brand for the city, a response to criticisms that London lacked a clear tourism identity (Tyler 2009). Despite responsibility for tourism shifting between different organisations, criticisms of the way tourism is managed have remained consistent. One recurring complaint is that tourism is regarded merely as an economic activity, rather than one that affects and is affected by wider London's socio-cultural landscape. As responsibility for tourism at the city-wide level is now allocated to a destination marketing agency, it seems unlikely that this problem will be addressed by the current regime.

Academic Attention

It is surprising that in academic literature, London's tourism sector and the multiple issues associated with it have received relatively little attention. During his time at Birkbeck College, Andrew Church (working with different co-authors) published some useful work on business tourism (Church and Bull 2001) and labour issues (Church and Frost 2004). Later, Robert Maitland's work (featured in this volume) on 'off the beaten track' tourism in London and related research on cultural tourism by Steve Shaw (Shaw and Macleod 2000; Shaw et al. 2004; Shaw 2008; 2011) made a significant contribution to our understanding of tourism beyond the obvious (centrally located) attractions. There has also been some attention to the sustainability of the tourism sector (Knowles et al. 1999) and post-disaster recovery (Ladkin et al. 2008), plus some fascinating work on historical representations of London as a tourist destination by David Gilbert (Gilbert 1999; Gilbert and Hancock 2006; Driver and Gilbert 1998). Nevertheless, tourism in the UK capital is not well covered by the academic literature, especially if one considers the contemporary significance of London as one of the world's most visited destinations. In her recent work on planning and managing tourism in London (2015; 2016; 2017), Cristina Maxim (2017, 1) recognises that 'despite the important role tourism plays in the economy of the city, there is limited research on the development of this activity in the capital'. This book aims to fill this conspicuous gap in the literature regarding this imbalance by providing a book dedicated to the contemporary tourism sector and its expansion.

The Structure and Content of the Book

Destination London examines how tourism has extended into parts of London not normally regarded as visitor destinations. As Biddulph (2007) argues, spatially oriented studies of tourism have always been fascinated by back regions and the ways these are commodified, and many of the chapters here shed further light on this type of expansion. The book begins with Robert Maitland's review of the spread of tourism into non-central areas, including the city's suburbs (Chapter 2). This chapter provides a good introduction to the ideas and trends that underpin the shift towards a 'new urban tourism', where tourists penetrate further into the city in a search for more distinctive and more authentic districts. In contemporary London, various factors are responsible for tourism's spatial expansion, but the rise of peer to peer accommodation and the sharing economy seem pivotal. These trends and their impacts on the city are discussed by Clare Inkson in Chapter 3. Other spatial expansions of tourism are facilitated by new developments in the urban periphery. Tourism in London was originally driven by the railways and the grand stations that were built in the nineteenth century, but in the contemporary era, it is airports on the edge of London that provide the city's gateways. These sites are the focus of Anne Graham's work in Chapter 4. In London, rapidly expanding aerotropoli can be understood as part of the wider city destination and as urban destinations in their own right. Sports stadiums are another important feature of London's non-central districts, and their role in driving tourism in peripheral districts is discussed by Claire Humphreys in Chapter 5. Many stadiums have been constructed or reconstructed in recent years and one key design principle is to satisfy growing interest from visitors.

Chapters 2 to 5 essentially focus on the spatial expansion of tourism, helping to explain why tourists are visiting areas outside Westminster, The City of London, Camden, and Kensington and Chelsea. Subsequent chapters (6 to 8) focus on more subtle extensions of tourism in more central areas – into the air, onto aquatic territory and through hidden worlds. These chapters explain how London's tourism territory has extended through the provision of new ways of consuming London. Rather than representing new products in new districts, they are essentially ways of consuming central districts from a different perspective. In Chapter 6 Andrew Smith examines the new ways London can be consumed from above – highlighting the recent provision of dynamic experiences rather than merely visual ones. The subsequent Chapter (7) by Claudia Dolezal and Jayni Gudka highlights a different form of expansion: one that involves opening up secret worlds and alternative interpretations – by offering tours led by homeless people. Simon Curtis then highlights the way that the River Thames has been opened up to provide new open space and new vantage points for tourists (Chapter 8).

Chapters 9 and 10 focus more on events. The work by Adam Eldridge and Ilaria Pappalepore on tourism in the winter season – and at night – means the book also addresses how tourism expands temporally. Andrew Smith then examines the ways London's neighbourhood parks are being integrated into the visitor economy through their transformation into event venues. In an era of neoliberal austerity, resources previously regarded as local amenities are revalorised as eventscapes, thus expanding the reach of tourism into new territory. The core themes of the book and their implications are discussed in the concluding Chapter (11), alongside recommendations for future work.

References

Biddulph, Robin. 2017. 'Tourist Territorialisation and Geographies of Opportunity at the Edges of Mass Destinations' *Tourism Geographies*, 19(1), 27–43.

Bull, Paul and Andrew Church. 2001. 'Understanding Urban Tourism: London in the Early 1990s'. *International Journal of Tourism Research*, 3(2), 141–150.

Butler, Tim. 1997. *Gentrification and the Middle Classes*. London: Ashgate Publishing.

Church, Andrew and Martin Frost. 2004. 'Tourism, the Global City and the Labour Market in London'. *Tourism Geographies*, 6(2), 208–228.

Crookston, Peter. 1978. 'Introduction'. In Peter Crookston, ed. *Village London. The Observer's Guide to the REAL London*. London: Arrow.

Driver, Felix and David Gilbert. 1998. 'Heart of Empire? Landscape, Space and Performance in Imperial London' *Environment and Planning D: Society and Space*, 16(1), 11–28.

Gilbert, David. 1999. 'London in all its Glory—or how to enjoy London: Guidebook Representations of Imperial London'. *Journal of Historical Geography*, 25(3), 279–297.

Gilbert, David. 2006. 'The Youngest Legend in History: Cultures of Consumption and the Mythologies of Swinging London'. *The London Journal*, 31(1)1–14.

Gilbert, David and Claire Hancock. 2006. 'New York City and the Transatlantic Imagination: French and English Tourism and the Spectacle of the Modern Metropolis, 1893 –1939'. *Journal of Urban History*, 33(1), 77–107.

GLA. 2015. *Daytime Population of London 2014*. Available at https://data.london.gov.uk/apps_and_analysis/daytime-population-of-london-2014/ (accessed 7 December 2018).

Knowles, Tim. et al. 1999. 'The Development of Environmental Initiatives in Tourism: Responses from the London Hotel Sector'. *International Journal of Tourism Research*, 1(4), 255–265.

Ladkin, Adele. et al. 2008. 'London Tourism: A "Post-Disaster" Marketing Response' *Journal of Travel and Tourism Marketing*, 23(2–4), 95–111.

Lim, Sonia and Frederic Bouchon. 2017. 'Blending in for a Life Less Ordinary? Off the Beaten Track Tourism Experiences in the Global City'. *Geoforum*, 86, 13–15.

London and Partners. 2017. *A Tourism Vision for London*. Available at https://files.londonandpartners.com/l-and-p/assets/london_tourism_vision_aug_2017.pdf (accessed 7 December 2018).

Maxim, Cristina. 2015. 'Drivers of Success in Implementing Sustainable Tourism Policies in Urban Areas'. *Tourism Planning and Development*, 12(1), 37–47.

Maxim, Cristina. 2016. 'Sustainable Tourism Implementation in Urban Areas: A Case Study of London'. *Journal of Sustainable Tourism*, 24(7), 971–989.

Maxim, Cristina. 2017. 'Challenges Faced by World Tourism Cities–London's Perspective'. *Current Issues in Tourism*, 22(9), 1–20.

Novy, Johannes. 2018. '"Destination" Berlin revisited. From (New) Tourism towards a Pentagon of Mobility and Place Consumption'. *Tourism Geographies*, 20(3), 418–442.

Polley, Martin. 2012. *The British Olympics: Britain's Olympic Heritage 1612–2012*. Swindon: English Heritage.

Shaw, Stephen. 2008. 'Hosting a Sustainable Visitor Economy: Lessons from the Regeneration of London's' Banglatown'. *Journal of Urban Regeneration and Renewal*, 1(3), 275–285.

Shaw, Stephen. 2011. 'Marketing Ethnoscapes as Spaces of Consumption: "Banglatown—London's Curry Capital"' *Journal of Town and City Management*, 1(4), 381–395.

Shaw, Stephen and Nicola MacLeod. 2000. 'Creativity and Conflict: Cultural Tourism in London's City Fringe'. *Tourism Culture and Communication*, 2(3), 165–175.

Shaw, Stephen, Bagwell, Susan, and Joanna Karmowska. 2004. 'Ethnoscapes as Spectacle: Reimaging Multicultural Districts as New Destinations for Leisure and Tourism Consumption'. *Urban Studies*, 41(10), 1983–2000.

Stock, Mathis. 2007. 'European Cities: Towards a Recreational Turn?' *Hagar. Studies in Culture, Polity and Identities*, 7(1), 115–134.

Tyler, Duncan. 2009. 'London's Tourism: Is a Purely Economic Approach Appropriate?' *Local Economy*, 24(5), 415–425.

CHAPTER 2

Extending the Frontiers of City Tourism: Suburbs and the Real London

Robert Maitland

Introduction

Two of the grand themes in tourism research and writing are growth – of tourist numbers, of destinations and of the scope of the visitor economy; and authenticity – the search for a tourist experience that seems 'real'. This chapter looks at the interaction between the very rapid growth in London's international visitor numbers, the city's changing economy and places, and tourists' concern with authenticity. It draws upon the University of Westminster's work on tourism in London and other World Tourism Cities (WTCs), which has shown that many visitors seek the 'real' city and that synergies between tourists and residents are important in reconfiguring, reimagining and reimaging places within the city (see for example Maitland (2007; 2010; 2014), Maitland and Newman (2004; 2009), Pappalepore et al. (2010; 2014), Cherifi et al. (2014)). In WTCs, tourism now thrives in once unfashionable areas of the inner city (for example, Brooklyn, New York City; Hoxton, London; Kreuzberg, Berlin (Maitland and Newman 2009)), and plays an important and synergistic role in the new economy of the inner city (Hutton 2009). But as development

---

How to cite this book chapter:
Maitland, R. 2019. Extending the Frontiers of City Tourism: Suburbs and the Real London. In: Smith, A. and Graham, A. (eds.) *Destination London: The Expansion of the Visitor Economy.* Pp. 15–35. London: University of Westminster Press. DOI: https://doi.org/10.16997/book35.b. License: CC-BY-NC-ND

pressures and tourism numbers increase, areas that were previously off the beaten track become incorporated into recognised tourism circuits and lose their distinctiveness. This means that 'urban explorers' (Maitland 2007) must look further afield in their search for the 'real' places where they feel they can get 'backstage' (Goffman 1959).

Swift and largely unplanned changes to London and in its population, together with shifting views about what constitutes 'the tourist', complicate the idea of a 'real city' that can be 'discovered' by visitors. So, tourists in search of the real city may now have to look further off the beaten track – into the suburbs. At present, suburbs seem unlikely new tourist destinations. However, place images can change, sometimes radically. Twenty-five years ago, attempts to attract visitors to inner city areas in the USA and the UK were novel and often derided (see, for example, Beioley et al. 1990); yet such areas now constitute London's hippest destinations.

This chapter draws on evidence and ideas from the University of Westminster's research on tourists' attempts to get off the beaten track in London and in other WTCs. It considers how far suburban areas can meet the demands previously satisfied by areas in the inner city, and whether their associated images and imaginaries can change as radically. The focus is London, though the ideas may prove applicable elsewhere. The chapter begins with consideration of why off the beaten track areas appeal to visitors, and examines the rapid changes in London that are shrinking what tourists have seen as the 'real city'. The qualities that constitute the 'real city' for visitors are identified, and the work assesses how far those qualities can also be found in the suburbs. The chapter concludes with an overview of the potential of suburban areas for tourism, emphasising that negative image and imaginaries are crucial obstacles preventing the extension of tourism into suburbia.

Soft Tourism in the World City

The search for (lost) authenticity and a desire to get 'backstage' to discover 'real' places is a long established, though contested, theme in narratives of tourist practices and experiences (MacCannell 1999; Pearce and Moscardo 1986). 'Getting off the beaten track' has been more strongly associated with backpackers exploring exotic (to them) countries far from home. However, going 'off-piste' has become increasingly important to many city visitors, especially in WTCs, with their high-profile global brands and their capacity to generate new tourism areas. For some tourists, exploring the city and getting off the beaten track is at the heart of their visit, but for many more it is an important element in their overall experience of the city. They want to 'see the sights' and do some of the things that they know to be 'touristy' – yet also want to spend some time experiencing the 'real city' (Maitland and Newman 2009). The desire to be an urban explorer, for at least some of the time, stems from the increase in

London's visitor numbers, and changes in tourists' characteristics and preferences. There is a long-established growth in the number of international visits to London (see Chapter 1). Numbers increased from 14.7 million in 2010 to 15.5 million in 2012, 17.4 million in 2014, and 19.5 million in 2016. In June 2017, numbers were up 7 per cent on 2016 (Visit Britain 2017). There have been short-term variations, but growth has continued through exchange rate fluctuations, fears of war, terrorist incidents, and after the decision to leave the EU. This relentless rise has obvious consequences. Pressures on traditional tourist hotspots intensify, and affect the visitor experience. For the industry, there are stronger commercial incentives to produce a commodified tourist-scape, whilst visitors respond as they seek out places that seem to them less commercialised. At the same time, decades of uninterrupted growth in tourism have affected the tourists themselves. Many more are now frequent and experienced travellers who have already 'seen the sites' – both literally as they return frequently to cities like London, or metaphorically, because they have travelled extensively and have little desire to visit more 'top attractions'. 'Real London' rather than 'Brand London' provokes their interest and adds to their cultural capital. Finally, more visitors are 'connected tourists' (Maitland 2014), people who know the city well because they previously lived, worked or studied there or are connected to it by the friends and relatives they have come to visit or the work colleagues they meet when they come to the city on business. Connections mean these tourists have ready access to the 'backstage' places, and perhaps a strong motivation to continue to explore the city they used to live in, or to experience the city life of their friends, relatives or colleagues. (See Chapter 3 for a discussion of the impact of Airbnb).

Moreover, for experienced and connected visitors, the focus of city tourism is shifting. It is moving away from relying principally on exploiting tangible resources like historic buildings or museums and galleries towards a concern with intangible resources like lifestyle and image. That means that 'having' a holiday, or 'doing' the sights has less appeal than 'becoming' different through the effects of the tourist experience (Richards and Wilson 2007). For Andersson Cederholm (2009) 'being' is an emerging tourism value: being with oneself, in a contemplative fashion; being with co-tourists, especially those with shared values and interests; and being with local people – an essential element in experiencing place. At the same time, it has become increasingly difficult to isolate and separate tourists and touristic practices as tourism comes to be seen as simply one of a suite of mobilities (Hannam 2009), and touristic practices overlap with those of city residents (Franklin and Crang 2001).

The result is that many tourists are urban explorers for at least some of their visit. We can see this as 'soft tourism' whereby 'tourists albeit temporarily, "embed" themselves ... and experience locally distinct cultural activities, products and environments' so that they can integrate themselves in the city as they 'occupy the same physical spaces and satisfy their existential and material needs in the same manner as members of the host society' (Oliver and Jenkins

2003, 296 and 297). In other words, urban explorers want to find distinctive places where they can integrate themselves in everyday life, and so experience the real. As Hall (2007, 1139–1140) says, 'Fakery occurs when the form of the physical or social object loses its integration with the everyday life of the place in which it is situated', whereas 'authenticity is born from everyday experiences and connections which are often serendipitous, not from things "out there". They cannot be manufactured through promotional and advertising deceit or the "experience economy"'. However, as London changes, it becomes more difficult to find the real city and enjoy soft tourism, the everyday and serendipity.

So urban explorers seek a soft tourism experience – which allows them to experience the real city by finding ways to embed themselves in it – by exposing themselves to serendipity and the everyday. However, changes in the city itself mean they need to be resourceful to do so.

Real London and Brand London

In one sense any place is authentic and real – it is as it is. But as Knox (2005, 3) points out, drawing on Heidegger, elements of the modern world – telecoms, technology, mass production, mass values – subvert the 'authenticity' of place so that 'city spaces become inauthentic and "placeless", a process that is, ironically, reinforced as people seek authenticity through professionally designed and commercially constructed spaces and places whose invented traditions, sanitised and simplified symbolism and commercialised heritage all make for convergence rather than spatial identity'. As Real London recedes, visitors' search for authenticity drives the growth of Brand London. We can see commercial spaces as attempts to satisfy visitors' demands for existential authenticity where the place conforms to the city of their imagination. Salazar (2013, 34) argues that imaginaries are 'socially transmitted representational assemblages that interact with people's personal imaginings and are used as meaning-making and world–shaping devices' and that exoticised imaginaries of otherness prompt tourism. Potential tourists imagine a 'paradisiacal environment where local landscape and population are to be consumed through observation, embodied sensation and imagination'. Such paradisiacal environments are not confined to clichés of white beaches and waving palms. Local landscapes and populations can be consumed in these ways in cities – by embedded tourists.

Imaginaries of cities are complex and, in some ways, contradictory. London is well known and well publicised, a carefully promoted global brand, and is undergoing radical and rapid change. Yet imaginaries of London are slow to change. Research on the images of London held by Czech non-visitors (Cherifi et al. 2014) show that images that would appear very old-fashioned can be stable and slow to change. There have been energetic attempts to refashion London's image – not least through the expensive staging of the 2012 Olympics. However, Visit London's (2015) advice to first time visitors featured just three main images: Buckingham Palace, Tower Bridge and Piccadilly Circus (along with a subsidiary image of visitors poring over a large paper map: very retro). The imaginary of heritage, history and royalty remains well supported.

Yet over the past 20 or so years, London has been changing radically and profoundly. As Kuper (2015) shows, it has risen to the top of global reputational league tables – constantly vying with New York City. He argues that three factors account for this. First, London is now a global rather than a national capital and attracts money and talent from across the world. Second, it has become more colourful – for example through renewed public spaces, spectacular architecture like Tate Modern or the London Eye, a renowned restaurant scene, street life, bars and cafes – and become more eventful: the Olympics are only the most obvious example. It has become more colourful, too because of its cosmopolitan population, attracted to London in part because now 'it is a place without a dominant national culture … to most foreigners London now looks like a place where you can self-actualise' (Kuper 2015, 3). A good place for being and becoming, then, but one in which the sense of place and sense of itself is blurred, complex and contradictory. Finally, and paradoxically, London offers stability – a long history, institutions that have evolved slowly, and sufficient political stability to attract global elites who want somewhere safe to keep their money and their family. Brexit will impact on all these trends, but thus far seems unlikely to change them profoundly.

This evolution has transformed places within London, most obviously through very rapid rises in property prices, seen by commentators as driving the working classes, lower middle classes and increasingly the professional middle classes from the central and inner city (see for example Minton 2017). This is what Erenhalt (2013) termed 'the great inversion' of a long-established pattern of poor inner cities and prosperous suburbs. Now, wealthy elites have moved back to the central and inner city, whilst the poor, the less well-off and migrants move to outer areas. Indeed, some once bustling parts of the most expensive areas of central London have become quiet, as more and more of the housing stock is acquired by foreign owners, who are frequently absent or see their property as a secure asset to be kept vacant – 'safety deposit boxes in the sky' as the former Chief Planner at the City of London put it (Rees 2015). However, processes of real estate speculation and gentrification have reached into formerly unfashionable areas throughout inner London. As Ehrenhalt notes, 'creatives' and hipsters colonise rundown areas, attracted by low property prices and the opportunity to display their love of 'edginess'. They are followed by bourgeois–bohemians (bobos) many of them foreigners. As gentrification proceeds, the wealthy move in. In 2012, London residential property worth £83 billion was bought for cash – by those working in the City financial district, and by rich foreigners seeking a safe and profitable investment (Goldfarb 2013). This process provides an urbanism that is attractively well manicured and may be aesthetically appealing – but one where the private realm displaces the public (in gated communities or commercial spaces to which public access is permissive, not an entitlement), and ideas of mixed communities are absent. The urban atmosphere may be appealing, but is placeless.

Tourism has played a significant role in transforming and reimaging run-down areas, with some tourists' urban preferences linking synergistically and seamlessly with those of some residents, and with tourism spending and tourist presence supporting the gentrification process (see for example Maitland and Newman 2009). However, super-gentrification and the profitability of new residential development is undermining the qualities that made the areas attractive, as rising real estate prices force out even long-established independent small businesses, restaurants and shops.

The very rapid increase in London's visitor numbers has helped drive the transformation of central and inner London, with areas that were once 'undiscovered' and off the beaten track increasingly drawn in to the commercialised tourist heart of the city. Although inevitably celebrated by much of the tourism industry, this process is problematic. As Bell and Welland (2007:2) commented, London is becoming as high-rise as New York City (see Chapter 6), and 'it can sometimes seem as though there is nobody over 30 on the streets and that a great experiment in mass immigration and assimilation is under way … in an effort to capture the flag from NY, London risks losing what makes it London. Of course, areas and places in a dynamic city change constantly. In the 1960s, 'Swinging London' saw the incorporation of once off the beaten track areas like Carnaby Street and the King's Road in a newly fashionable and vibrant commercial scene (Rycroft 2002). But recent changes in London have been of a different scale. Perhaps, as Goldfarb (2013, SR5) claims, 'the delicate social ecology that made possible London's transformation into a great world city over the last two decades is past the tipping point'. For 'hard' tourism, often first-time visitors in organised groups who want to see London's iconic sights this may not matter too much; indeed, the addition of new 'world class' developments may seem an advantage. However, those whose imaginaries are of a different London and who want a more integrated soft tourism will need to work harder to search out the 'real London'.

Cool and Convivial: Getting off the Beaten Track

Research by Westminster academics on London and other World Tourism Cities has shown that some tourists want to get away from popular hotspots to places that seem off the beaten track. In London, the research has included visitor surveys with almost 400 respondents, and lengthy semi-structured interviews with a total of more than 200 interviewees, at non-central locations in the inner city (for example Islington, Bankside, Spitalfields, Hoxton, London Fields and Deptford); the research has been fully reported elsewhere (for example Maitland and Newman (2004; 2009), Maitland (2008), Pappalepore et al. (2011; 2014)). This research draws out three aspects of their experience that allow urban explorers to get off the beaten track and feel they can embed themselves in the city. They are the combination of morphology and consumption landscape; image and imagined geography; and experiencing everyday life. Following Lefebvre (1991) and drawing on Collis, Felton and Graham (2010, 1050) in their discussion of suburbs, we can see these as the three elements that go to constitute place: the objective material space; the way space is imagined and represented; and how it is experienced.

The morphology of the areas is crucial for visitors, and they frequently describe and comment on buildings and urban form in detail. The areas visited are characteristically formerly industrial, working class and under-privileged, often with a strong representation of ethnic minority populations. Their urban form seems organic and unplanned, is at comparatively high density, and has intricate street patterns and buildings of a human scale. Visitors contrast this with tourist hotspots, seen as having monumental architecture and layout, or commercialised environments that seemed planned for visitors. Unlike monumental or carefully choreographed commercial environments, such places offer simultaneous rather than successive arrangements of spatial elements (Gospodini 2001), meaning that visitors have many options and choices in how they move around them. They are, in other words, easily and temptingly explorable. Indeed, a minority of visitors specifically commented on the pleasure of 'getting lost' – whilst knowing that they could and would regain their bearings. This intricate urban form contains a mix of land-uses and seems to have more independent businesses, often in the creative sector – arts, fashion, food, craft beers and so on – providing an attractive landscape of consumption. Branches of well-known national and international chains are comparatively rare. These qualities of the objective material space contribute to places that are distinctive and have a buzz.

The image or imagined geography of space intersects with this objective material space, and contributes to fulfilling the expectations many visitors have of the 'real London'. In these multi-purpose and heterogeneous spaces 'with blurred boundaries … a wide range of activities and people co-exist. Tourist facilities coincide with businesses, public and private institutions and domestic housing, and tourists mingle with locals, including touts … heterogeneous tourist spaces provide stages where transitional identities may be performed alongside the everyday actions of residents, passers-by and workers' (Edensor, 2000, 64). Novy and Huning (2009, 87) point out – when discussing Berlin – that 'particularly edgy, transitional and allegedly authentic urban settings such as industrial and warehouse districts, ethnic or immigrant enclaves and other neighborhoods where people on the margins of urban society live and work are today part of a growing number of tourists' travel itineraries … Former no-go-areas have been turned into desired travel destinations, as their "authenticity", the alternative lifestyles of their residents and their different tangible and intangible cultural resources – music, art, history, traditions, the aesthetic of their built environment etc. – became attractive for outsiders'. This links to Nancarrow et al.'s (2001) discussion of what constitutes 'cool'. For them it revolves around a search for the authentic and a valuing of insider knowledge about trends and consumption patterns outside the mainstream – a form of cultural capital. As Bovone (2005, 377) suggests, 'a fashionable quarter is one where urban lifestyles and production … are initiated before elsewhere'. Off the beaten track areas can satisfy the demand for a real London hidden from the mainstream, known only to insiders, and in some ways responding to a nostalgic desire for a city with an intimate villagey built environment and a consumption landscape of trend-setting independents, removed from homogenising global businesses. These places are imagined and represented as distinctive, since they have emerged organically through micro interactions in the market, and have not been planned as spaces for consumption by developers or public authorities. They are yet to be 'commercially appropriated' (Neill 2001) and their rundown origins offer 'grit as glamour' (Lloyd 2000) where visitors can experience 'safe danger'.

Subsequent work (Pappalepore et al. 2011; 2014) has investigated the role creative clusters play in the development and experience of tourism off the beaten track. We found that concentrations of creative industries provide visitors with opportunities for consumption and for the accumulation of cultural capital, drawing on and exploiting the presence of creative producers and other creative visitors, who are themselves perceived as an attraction. In such creative tourism areas, these elements combine with others we have already discussed – a particular morphology, and the opportunity to embed oneself in the everyday life of the city – to produce places that visitors see as real, with a bohemian atmosphere and cool image. Whilst we identified several varieties of practice (Edensor 2000) in the ways that visitors engaged with the areas, for most tourists, the sense that they were getting away from the mainstream was central to the appeal of the areas.

Places that are distinct from established, planned or commercially developed tourist bubbles offer the opportunity to experience the everyday life of the city, and mundane activities and routines become invested with interest and meaning. Quotidian activities like daily shopping, or people at work or in a café are interesting to observe, and confirm that these are not places planned for visitors. As one interviewee commented 'it doesn't feel artificial … you don't feel like you're in Disneyland'. Local people are key markers and signifiers that these are real places, and provide both confirmation of authenticity and a sense of the exotic. Another interviewee said 'tourist spots are always very generic, right, look at the places where tourists are in any city you feel like, oh, I'm just one of them and I'm just doing the typical tourist thing but if you, somehow, end up in the place where the locals go, it feels like a more authentic experience'. For others, Tesco, a mid-range supermarket, was 'one of our favourite places'. Of course, these tourists had not come to London specifically to spend time in Tesco. Rather, they wanted to integrate into the everyday life of the city, and the supermarket allowed them to do so, to mingle and observe local people, see what they bought, and to participate in quotidian life. This opportunity was valued: 'it's more authentic and fun, because local people and tourists, they also mix. Here, you are not treated as a tourist' (Maitland 2008). A convivial relationship between tourists and locals seems an essential element in the experience of everyday life. In short, the tourist gaze of the outsider creates the exotic from ordinary life: the everyday is not simply ordinary – 'rather it is the site that contains the extraordinary within the ordinary if one is prepared to look' (Till 2009, 139). However, we should bear in mind that 'local people', from the tourist perspective, mean simply non-tourists. High levels of migration and rapid churn in the population of London's neighbourhoods means that meeting with truly 'local' people is comparatively uncommon – if by that we mean those born and bred in the area or who are long term residents.

The desire of some visitors to experience what they regard as the real city by getting off the beaten track seems clear. The 'real' derives from a combination of objective space, the experience of space, and imagined geography. However, getting off the beaten track in inner London has become increasingly difficult as booming property markets and rapidly increasing populations accelerate the rate of commercial appropriation and squeeze areas once seen as outside the mainstream and offering the possibility of experiencing everyday life. For those whose imaginaries are of a London of explorable places with interesting vernacular architecture, varied landscapes of consumption and little evidence of hard tourism, many gentrified areas will retain their appeal. We found that for some visitors, upscale neighbourhoods like Islington can feel like the real London, though they are increasingly dominated by global elites (Butler 2007). Areas like Spitalfields have continued appeal to many visitors, but now seem mainstream to those in search of cool places. They certainly no longer constitute Novy and Huning's (2009, 87) 'edgy, transitional and allegedly authentic urban settings'. Relentless development pressures in inner London have meant that little is now off the beaten track. Perhaps those in search of cool and convivial places and the real city should look to the suburbs.

Finding the Real City in the Suburbs

Hinchcliffe (2005, 2) points out that 'the literature on suburbs is extensive, and yet the subject always seems elusive. For some the suburb is a geographical space, for others a cultural form ... for others a state of mind'. In other words, different commentators put different emphasis on the components of suburbs: their objective material space, imagined geography and experience of the everyday. This helps account for one of the difficulties of discussing suburbs and their potential appeal – avoiding 'the dangers of over-generalizing about cities and suburbs' (Phelps 2012, 259). It is especially important to avoid the illusion that the city's centre and periphery developed independently of one another. In reality, whilst suburbs have very different characters, they

Figure 2.1: Suburban Otherness: The Imagined Geography of Suburbs (Photo: John Maltby/RIBA Collections).

cannot be regarded in isolation from either the central city or its surroundings (Hinchcliffe 2005); rather they form part of a complex urban region. Perhaps this means that traditional distinctions are now meaningless. For Lang and Knox (2009), 'the city' and 'suburb' are 'zombie categories', irrelevant in a contemporary context.

London's suburbs are disparate and varied in their character (Phelps 2012). The Victorian development of London saw the construction of suburbs in what has since become inner London, whilst the outer suburban areas were constructed mainly in the twentieth century. In both eras, suburbs frequently grew, as had other parts of the city, from a pre-existing village nucleus. Some were predominantly residential but others were substantially industrial (e.g. Wembley and Willesden), and others had a mix of small businesses and housing (e.g. Acton). The high amenity inner and outer suburbs (Camberwell, Hammersmith, Putney, Ealing, for example) provided for those moving in search of more personal independence and freedom; they helped create a market for arts and crafts products and provided a home for new colleges providing arts education and training (Phelps 2012). Rather than there being a clear distinction between (inner) city and suburbs, we can see many shared qualities. The morphology of suburbs can echo many qualities of the inner city, with intricate street patterns stemming from village origins and complex patterns of land ownership. Nineteenth and twentieth-century suburbs mix housing with small industrial buildings capable of conversion to other uses – lofts, workshops, studios and so on – whilst some larger industrial buildings have been converted to residential loft apartments or re-used as performance spaces or complexes of studios and workshops. There are architectural and heritage attractions ranging from William Morris's Red House in Bexleyheath; Eltham Palace, a royal palace transformed in the 1930s with an art deco interior; to Lawn Road flats (the Isokon Building), an architecturally influential modernist building that became a centre of north London intellectual life. Indeed, clusters of creative industries are to be found in several parts of suburban London (Freeman 2009). Despite the familiar arguments of Jacobs (1961) and Florida (2005), creative industries and creative workers are not confined to the inner city – they happily locate in suburban areas (Flew 2012; Collis et al. 2013). Indeed, it is argued that the 'bourgeois utopia' of high amenity suburbs are being reconstituted as locations for emerging small businesses including in professional and creative sectors, as urban businesses value proximity to home along with public and private services, amenities and green space whilst retaining links to regional professional and industry networks (Phelps 2012, 266).

And suburbs are of course pre-eminently the scene of everyday life, since they are 'the principal residential environment for the majority of the population' (Whitehand and Carr 2001, 182). Indeed, as London transmutes into a global capital with central and inner areas colonised by global elites, the suburbs are increasingly where 'the locals' are to be found – if by that we mean those for whom the city is their permanent and long-standing residence. London has transformed into a mega city of global migrants in which the majority (55 per cent) of the population is not White British. Much of that majority live in outer London – the suburbs: in 2011, half the black population and two thirds of the Asian population lived in outer London. Poles, Somalis, Afghans and Ghanaians live in places like Beckton, Ilford, Edmonton, Catford and Harlesden (Judah 2016). They bring with them culture, events, shops and restaurants that visitors in search of the exotic of the everyday may relish. Perhaps there are emerging similarities with Paris, a city in which the Boulevard Périphérique has long marked a clear divide between 'city' and 'suburbs'– the banlieues. Maspero (1994, 16, quoted in Phelps 2012) claims that it is in the banlieues that real, authentic life is to be found:

> 'where did they all go? To the outskirts. To the suburbs. Paris has become a business hypermarket and a cultural Disneyland … And didn't that mean the true centre was now "all round"?'

So, in terms of morphology, of objective material space, London's suburbs have many of the qualities of the inner city. Moreover, they are the real city, in which visitors who want to experience the exotic of the everyday can find it. And suburbs already receive many visitors. People visiting friends and relations go to where their friends and relatives are to be found – frequently in the suburbs. Meanwhile enterprises like Airbnb make it easier to let rooms to visitors in unfamiliar areas (see Chapter 3) whilst rising property prices in central and inner London encourage budget hotels in outer areas. Yet we hear little of the appeal of suburbs for tourism, or the possibility that the well-established expansion of areas that tourists visit will continue outwards. This apparent paradox is resolved when we consider that the imagined geography of suburbs is relentlessly negative – and has increasingly diverged from reality (Collis et al. 2010). Any suggestion that suburbs may be attractive to visitors – or even cool – has run up against an apparently entrenched view that they are 'maligned … connoted an inferior form of city … an easy [insulting] epithet … shorthand for hypocrisy and superficiality' not least because limited academic attention has meant our 'understanding [has been] … restricted to an odd mix of cliché and dated pop culture' (Kirby and Modarres 2010, 65).

This negative imagined geography of suburbs has been constructed from academic and professional discourse and from high and popular culture. Ideas of a suburban dystopia, destructive of both city and countryside, can be traced in England at least from the work of Ruskin in the later nineteenth century, and a key purpose of the planning system that emerged in the UK with the 1947 Town and Country Planning Act was to manage suburban development and prevent sprawl. However, there was always more to this than an attempt to manage land-use patterns, and attitudes were inflected with a criticism of the imagined culture and politics of suburbs. Ian Nairn (1955, 365) in a provocatively polemical contribution saw suburbs as:

> the creeping mildew that already circumscribes all our towns. This death by slow decay is called subtopia … the world of low-density mess.

Whitehand and Carr (2001) point to the strong professional disdain of the suburbs by architects and planners, perhaps because of a built form that focuses on the individuality of single-family dwellings rather than the collectivist form of the Georgians or Modernists. They see this as accompanied by an intellectual scorn for the suburbs, presented as places inhabited by the undereducated lower middle classes, who are portrayed as conservative and status conscious. More recently, Florida's (2005) influential work on the creative class explicitly contrasts the bohemian enclaves of a dense inner city with the sprawl and (alleged) lack of creativity of the suburbs. So, suburbs come to be 'mythologised as places that exist somewhere else and are inhabited by people unlike ourselves' (Vaughan et al. 2009, 9): suburbanites are 'the Other' (Figure 2.1). Phelps (2012, 268) sees this as intellectual snobbery, and comments that the 'privileging of the city within academic and policy discourse may simply be the latest incarnation of "suburb bashing" by elites' and reflect 'imaginings of their own social worth'. Yet a sense of 'suburban otherness' may give a clue to what may attract tourists in search of the real.

This professional and academic disdain runs alongside similar cultural attitudes. In the 1890s, the satirical magazine *Punch* parodied the inhabitants of the new suburbs (now absorbed into inner London), most famously Mr and Mrs Pooter in *Diary of a Nobody*. Aspirant suburbanites, 'working hard to improve their economic and domestic security and claim the right to personal meaning for their lives' were sneered at by the established middle classes (Hapgood 2000, 40). George Orwell's attitudes to suburbs and their inhabitants were complex but in *Coming Up for Air* are overwhelmingly negative. 'You know how these streets fester over the inner-outer suburbs. Always the same. Long, long rows of little semi-detached houses' (Orwell 1939, 54). (The description of featureless and alienating places, inflected with the image of disease or infestation, is echoed by Nairn, above, and is common to much commentary on the suburbs). When George Bowling, the protagonist, tries to escape by returning to the nostalgically remembered England of his boyhood, he finds the village he grew up in has itself been engulfed in suburban development. The suburb stands for the inauthenticity and falseness of modern life, encapsulated when Bowling orders a frankfurter in the pub, and finds it tastes of fish. Literature and drama have retained this perspective on the suburbs. The critic Q.D. Leavis (1965, quoted in Webster 2000, 4) was especially disdainful: 'suburban culture … has no fine rhythms to draw upon and is not serious … it is not only formed to convey merely crude states of mind but it is destructive of any fineness'. The same attitudes can be seen in more popular work. Mike Leigh's 1970s stage and TV play *Abigail's Party* remains critically celebrated, but its disdainful view of suburban pretension shares attitudes with many largely forgotten suburban sitcoms like *Bless This House, George and Mildred, The Fall and Rise of Reginald Perrin* or *The Good Life*. Their 'sustained and popular indictment of suburbia … a monotonous world of lonely, frustrated housewives and henpecked husbands might have been scripted by upper-class intellectuals of the inter-war years'; although suburbs 'represented a dream come true for millions of ordinary families … intellectuals denounced their small-mindedness' (Sandbrook 2010, 331). The satirical magazine *Private Eye* (founded and edited by Oxbridge-educated public schoolboys) continues to present the suburb of Neasden as 'the symbol of everything base, boring and banal … where romance and imagination came to die' (Sandbrook 2010, 330).

We could go on. As Webster (2000, 4) says:

> There is a remarkable degree of consistency indeed uniformity in external perspectives on suburbia. The defining characteristics whether viewed from the country or the city tend to be reducible to unimaginative conformist design and behaviour determined by imitation rather than originality; a lack of individuality combined with excessive social homogeneity; spatially cramped and confined conditions and a neglect for, or undermining of, traditional values

He goes on to point out that some commentators are much more nuanced and interested in exploring the contradictions of suburbia. Some writing about suburbia displays a fondness, even nostalgia, for example the work of John Betjeman, who coined the term Metro-land for the suburbs along the Metropolitan Line from Baker Street, stretching North West out of the capital, or of Philip Larkin who 'definitely thought suburbanites and small towners live more authentic lives' (Harris 2010); or displays tensions and contradictions (the work of Hanif Kureshi or Nick Hornby, for example). And since the 1960s there has been a strand of English music that gently mocks the suburbs but values them – The Kinks' 'Shangri-La' and 'Muswell Hillbillies', The Jam's 'Tales from the Riverbank' as well as 'Wasteland'.

So, the relentless negativity of the imaginaries of suburbs is only part of the story; there is a fondness. But overwhelmingly, the portrayal of suburbs by academic and professional commentators is negative, despite some notable exceptions such as the early Willmott and Young (1960) study of the rich community life in a London suburb, and has been reinforced in popular and high culture. It is difficult not to see strong class elements here, as economic and cultural elites disdain the aspiration of lower middle and working classes, and scorn their attempts to change their class position. And this echoes familiar stereotypes within tourism: the superiority of the 'traveller' with high levels of economic, social and cultural capital to the plebeian mass tourist. Despite countervailing and revisionist views, that is a position that is hard to change. As Salazar (2013, 36) points out, tourism imaginaries can be immobile: 'in some destinations tourism imaginaries are so firmly established and all-encompassing that they are difficult to escape'. Yet, this is an imaginary that diverges from objective reality, and is out of date. Many suburbs share the morphological qualities of much of the inner city, and are home to creative industries and those who work in them. In contemporary London, the juxtaposition of boring, conformist, inauthentic and standardised suburbs with an inner and central city that is vibrant and authentic is not only an inaccurate and unflattering portrayal of suburbs – it is an inaccurate and far too flattering portrayal of the inner and central city. If London is turning into a 'mass gated community of the world's richest people' (Kuper 2015, 5), then the suburbs are the place to go for visitors who want to get off that beaten track and experience the real life of the city.

Conclusions: The New Real London

This chapter has drawn on extensive research in London and other WTCs to argue that many tourists want to get off the beaten track to discover the real city; that economic change, real estate development and rapid growth in visitor numbers mean that few parts of London's inner city can now be seen as 'undiscovered'; and that suburbs can offer the qualities that urban explorers seek.

Growth in the numbers of tourists who are experienced travellers, often connected to the city they visit, has combined with the desire to experience the real and authentic to drive some visitors to leave well-established tourist beats and seek out new areas. These places seem to offer a real experience through a combination of morphology, an imagined geography that is distinctive, and the opportunity to experience the everyday life of the city – where exoticism can be found in the everyday, and there is an opportunity to fit in rather than stand out, whilst mingling with co-tourists who seem cool. However, the radical changes that London is undergoing make getting off the beaten track in inner London much more difficult. Rapid and relentless property development driven by demographic change and London's role as a safety deposit box for foreign investors means that even the least fashionable and most run-down areas of inner London are becoming expensive. A previous development route which saw semi-derelict areas colonised by artists and creative industries seeking cheap space and developing in synergy with adventurous tourists and pioneer gentrifiers is now largely closed. At the same time, central and inner London is increasingly defined by transience (Goldfarb 2013) with the ultra-affluent more segregated and less committed to a city that is more of an asset store than a home. For locals and visitors who seek out areas that are authentic, for the opportunity to mingle with each other and co-tourists and pick up style tips, and who value the cultural capital and cool image to be derived from knowing about places outside the mainstream, inner London has less to offer.

One spatial consequence has been for artists, gentrifiers and curious visitors to look further afield – in some cases to other cities like Berlin. 'Eight years ago, Neuköln was considered to be dangerous, not even in the guidebooks. Now it's filled with tourists and expats. I'm part of a big exodus from east London to south-east London then to Berlin. The New Cross to Neuköln Express' (Kamradt 2015). Within London, they could look to the suburbs, now home to poorer residents and migrants, where property values are lower, and where everyday life is lived. Perhaps the Express could run from New Cross to New Eltham rather than to Neükoln. This would reflect a pattern that saw, for example, the Kings Road reimaged as fashionable in the 1960s and 1970s, Notting Hill in the 1980s and 1990s, and Shoreditch and Hoxton in the early twenty-first century, and provide opportunities for new creative and tourist areas well away from the mainstream, undiscovered and therefore cool. (It could of course also begin a process towards the problems of transience and overtourism now manifest in new tourism areas in inner London – for example loss of local identity, or transfer of housing assets into tourism accommodation. However, that prospect seems some way off). Yet the very notion of cool suburbs as a place attractive to tourists or hipsters still seems unlikely. As we have seen, this is despite similar morphology to formerly working-class inner London and it being the focus of the city's authentic everyday life. Rather, it is a consequence of a long established and relentlessly negative imagined geography that has made it almost literally impossible to imagine the suburbs as cool places, attractive to experienced travellers. Yet, there are reasons to think this may change.

The driving force of change is likely to be economy and demography as affluent incomers dominate inner areas, so that the suburbs and hinterland seem to have more to offer. But the very qualities that have made suburbs such objects of contempt may paradoxically build their attraction. If suburbs are home to 'the Other', then that in itself offers an exotic appeal for urban explorers. Webster (2000) sees the suburbs as liminal and ambivalent – not in the city, yet not outside it; not working class yet not upper class. Critics read this as superficiality and depthlessness – but the absence of a strong set of narratives and profound cultural signifiers could be seen as a strength. Wynn (2010) argues that the stuff of everyday experience, the free resources of culture, history and place, can be transformed into something meaningful – a process he terms 'urban alchemy'. In this process visitors use their experiences to create their own imaginaries and their own narratives of the city, drawing on everyday life and interactions with local people – both readily available in the suburbs. Suburbs are places where the everyday life of the city goes on, but which do not carry strong historical or cultural narratives – provided one can get away from a disdain of all things suburban. They are more malleable for the visitor, so that individual stories can be constructed; their otherness can be read as edgy, authentic and exotic. They can appeal to those 'tactical tourists' who 'look for places where they decide for themselves what they see and experience' and reject the 'specifically targeted strategies of the tourist industry' (Wolfram and Maier 2013, 362 and 365).

The growth of tourism in the outer city seems plausible, although I do not expect an immediate rush to the suburbs. It will be driven by the interplay of market forces and development opportunities with the desire of some tourists to escape places that have been commercially appropriated, as it was in off the beaten track areas in the inner city. Tourism developers and marketers will be involved, especially as the New London Plan (Mayor of London 2018) promotes densification and mixed-use development, often including hotels, in suburban hubs. However, their roles are limited – partly because their ability to intervene in development is circumscribed, partly because overt marketing of areas inevitably makes them mainstream. What would be helpful would be support for research. Currently there is almost no empirical work on how tourism is developing in the suburbs, how many visitors are involved, how far they explore the areas around their accommodation and whether processes are in fact comparable with those we have seen in the inner city. Tourism in the inner city was derided in the 1980s but is now integral to what London offers.

Perhaps in future, a visit to the cool suburbs will be equally essential – but we need more research before we can say so with confidence.

References

Andersson Cederholm, Erika. 2009. 'Being-with-Others: Framing the Ordinary as an Extraordinary Experience'. *ATLAS Annual Conference: Experiencing Difference – Changing Tourism and Tourists' Experiences* Aalborg, Denmark.

Beioley, Steve, Maitland, Robert, and David Roger Vaughan. 1990. *Tourism and the Inner City: An Evaluation of the Impact of Grant Assisted Projects.* London: HMSO.

Bell, Eugenia and Matt Weiland. 2007. 'London: the other New York'. *New York Magazine,* 25 October.

Butler, Tim. 2007. 'Gentrification'. In Nick Buck, Ian Gordon, Alan Harding and Ivan Turok, eds, *Changing Cities.* Basingstoke: Palgrave.

Bovone, Laura. 2005. 'Fashionable Quarters in the Post-Industrial City: the Ticinese of Milan'. *City and Community,* 4(4), 359–380.

Cherifi, Barbora et al. 2014. 'Destination Images of Non-Visitors'. *Annals of Tourism Research,* 49, 190–202.

Collis, Christy, Freebody, Simon, and Terry Flew. 2013. 'Seeing the Outer Suburbs: Addressing the Urban Bias in Creative Place Thinking'. *Regional Studies,* 47(2), 148–160.

Collis, Christy, Felton, Emma, and Phil Graham. 2010. 'Beyond the Inner City: Real and Imagined Places in Creative Place Policy and Practice'. *The Information Society,* 26(2), 104–112.

Davies, Fred. 1979. *Yearning for Yesterday: A Sociology of Nostalgia.* New York, NY: Free Press.

Edensor, Tim. 2000. 'Staging Tourism: Tourists as Performers'. *Annals of Tourism Research,* 27(2), 322–344.

Ehrenhalt, Alan. 2013. *The Great Inversion and the Future of the American City.* New York, NY: Alfred A. Knopf.

Flew, Terry. 2012. 'Creative Suburbia: Rethinking Urban Cultural Policy – the Australian Case'. *International Journal of Cultural Studies,* 15(3), 231–246.

Florida, Richard. 2005. *The Rise of the Creative Class.* New York, NY: Basic Books.

Freeman, Alan. 2009. *London's Creative Workforce: 2009 Update, Working Paper 40.* London: GLA Economics.

Franklin, Adrian and Mike Crang. 2001. 'The Trouble with Tourism and Travel Theory?'. *Tourist Studies,* 1(1), 5–22.

Goffman, Erving. 1959. *The Presentation of the Self in Everyday Life.* London: Penguin.

Goldfarb, Michael. 2013. 'London's Great Exodus'. *New York Times,* 12 October. Available at http://www.nytimes.com/2013/10/13/opinion/sunday/londons-great-exodus.html (accessed 29 June 2015).

Gospodini, Aspa. 2001. 'Urban Design, Urban Space Morphology, Urban Tourism: An Emerging New Paradigm Concerning their Relationship'. *European Planning Studies,* 9(7), 925–934.

Hall, C. Michael. 2007. 'Response to Yeoman et al.: The Fakery of the "Authentic Tourist"' *Tourism Management,* 28(4), 1139–1140.

Hannam, Kevin. 2009. 'The End of Tourism? Nomadology and the Mobilities Paradigm'. In John Tribe, ed, *Philosophical Issues in Tourism.* Bristol: Channel View Publications.

Hapgood, Lynne. 2000 'The New Suburbanites and Contested Class Identities in the London Suburbs 1880–1900'. In Roger Webster, ed., *Expanding Suburbia: Reviewing Suburban Narratives.* New York, NY: Berghahn Books.

Harris, John. 2010 'The Sound of the Suburbs and Literary Tradition'. *The Guardian,* 3 April. Available at http://www.theguardian.com/culture/2010/apr/03/suburbia-pop-betjeman-john-harris (accessed 8 February 2016).

Hinchcliffe, Tanis. 2005. 'Elusive Suburbs, Endless Variations'. *Journal of Urban History,* 31(6), 899–906.

Hutton, Thomas. 2009. 'Trajectories of the New Economy: Regeneration and Dislocation in the Inner City'. *Urban Studies,* 46 (5–6), 987–1001.

Jacobs, Jane. 1961. *The Death and Life of Great American Cities.* New York, NY: Random House.

Jenkins, Tim and Tove Oliver. 2001. *Integrated Tourism: A Conceptual Framework.* Deliverable 1, SPRITE Project.

Judah, Ben. 2016. *This is London: Life and Death in the World City.* London: Picador.

Kamradt, Johanna. 2015. 'Creative Young Brits are Quitting London for Affordable Berlin'. *The Guardian,* 1 August. Available at https://www.theguardian.com/world/2015/aug/01/creative-young-brits-quit-london-affordable-berlin (accessed 5 September 2017).

Kirby, Andrew and Ali Modarres. 2010. 'The Suburban Question: An Introduction'. *Cities,* 27(2), 65–67.

Knox, Paul. 2005. 'Creating Ordinary Places: Slow Cities in a Fast World'. *Journal of Urban Design,* 10(1), 1–11.

Kuper, Simon. 2015. 'This is Freedom? London's Reinvention'. In Ben Rogers, ed., *Soft Power.* London Essays, 1, Centre for London.

Lang, Robert and Paul Knox. 2009. 'The New Metropolis: Rethinking Megalopolis'. *Regional Studies,* 43(6), 789–802.

Larsen, Jonas. 2008. 'De-exoticising Tourist Travel: Everyday Life and Sociality on the Move'. *Leisure Studies,* 27(1), 21–34.

Lefebvre, Henri. 1991. *The Production of Space.* New York, NY: Wiley.

Lloyd, Richard. 2000. 'Grit as Glamour' cited in Dennis Judd 2003. 'Visitors and the Spatial Ecology of the City'. In Lily Hoffman, Susan Fainstein and Dennis Judd, eds. *Cities and Visitors*. Oxford: Blackwell.

MacCannell, Dean. 1999. *The Tourist: A New Theory of the Leisure Class,* Second Edition. London: Macmillan.

Maitland, Robert. 2007. 'Tourists, the Creative Class, and Distinctive Areas in Major Cities'. In Greg Richards and Julie Wilson, eds, *Tourism, Creativity and Development*. Abingdon: Routledge.

Maitland, Robert. 2008. 'Conviviality and Everyday Life: The Appeal of New Areas of London for Visitors'. *International Journal of Tourism Research,* 10(1), 15–25.

Maitland, Robert and Peter Newman. 2004. 'Developing Metropolitan Tourism on the Fringe of Central London'. *International Journal of Tourism Research,* 6(5), 339–348.

Maitland, Robert and Peter Newman. 2009. *World Tourism Cities: Developing Tourism off the Beaten Track*. Abingdon: Routledge.

Maitland, Robert and Andrew Smith. 2009. 'Tourism and the Aesthetics of the Built Environment'. In John Tribe, ed, *Philosophical Issues in Tourism*. Bristol: Channel View Publications.

Maitland, Robert. 2010. 'Everyday Life as a Creative Experience in Cities'. *International Journal of Culture, Tourism and Hospitality Research*. 4(3), 176–185.

Maitland, Robert. 2014. 'Urban Explorers: Looking for the Real (outer) London?'. In Greg Richards and Paolo Russo, eds, *Alternative and Creative Tourism*. Arnhem: ATLAS.

Mayor of London. 2018. *New London Plan – Consultation Draft*. Available at https://www.london.gov.uk/what-we-do/planning/london-plan/new-london-plan/draft-new-london-plan/ (accessed 12 October 2018).

Minton, Anna. 2017. *Big Capital*. London: Penguin.

Nairn, Ian. 1955. 'Outrage: On the Disfigurement of Town and Countryside'. *Architectural Review,* Special Edition.

Nancarrow, Clive, Nancarrow, Pamela, and Julie Page. 2001. 'An Analysis of the Concept of Cool and its Marketing Implications'. *Journal of Consumer Behaviour,* 1(4), 311–322.

Neill, William. 2001. 'Marketing the Urban Experience: Reflections on the Place of Fear in the Promotion Strategies of Belfast, Detroit and Berlin'. *Urban Studies,* 38(5–6), 815–828.

Novy, Johannes and Sandra Huning. 2009 'New Tourism (areas) in the New Berlin'. In Robert Maitland and Peter Newman, eds, *World Tourism Cities: Developing Tourism off the Beaten Track*. Abingdon: Routledge.

Orwell, George. 1939. *Coming Up for Air*. London: Victor Gallancz

Pappalepore, Ilaria, Maitland, Robert, and Andrew Smith. 2010. 'Exploring Urban Creativity: Visitor Experiences of Spitalfields, London'. *Tourism, Culture and Communication,* 10(3), 217–230.

Pappalepore, Ilaria, Maitland, Robert, and Andrew Smith. 2014. 'Prosuming Creative Urban Areas: Evidence from East London'. *Annals of Tourism Research*. 44, 227–240.

Pearce, Philip and Gianna Moscardo. 1986. 'The Concept of Authenticity in Tourist Experiences' *The Australian and New Zealand Journal of Sociology*, 22(10): 121–132.

Phelps, Nicholas. 2012. 'The Sub-Creative Economy of the Suburbs in Question'. *International Journal of Cultural Studies*, 15(3), 259–271.

Rees, Peter. 2015 'London needs Homes not Towers of Safety Deposit Boxes'. *The Guardian*, 25 January. Available at http://www.theguardian.com/commentisfree/2015/jan/25/planners-must-take-back-control-of-london (accessed 29 June 2017).

Richards, Greg and Julie Wilson. 2006. 'Developing Creativity in Tourist Experiences: A Solution to the Serial Reproduction of Culture?'. *Tourism Management*, 27(6), 1209–1223.

Richards, Greg and Julie Wilson. 2007. 'Tourism Development Trajectories: from Culture to Creativity?'. In Greg Richards and Julie Wilson, eds, *Tourism, Creativity and Development*. New York, NY: Routledge.

Rycroft, Simon. 2002. 'The Geographies of Swinging London'. *Journal of Historical Geography*, 28(4), 566–588.

Salazar, Noel. 2011. 'Tourism Imaginaries: A Conceptual Approach'. *Annals of Tourism Research*, 39(2), 863–882.

Salazar, Noel. 2013. 'The (Im)mobility of Tourism Imageries'. In Greg Richards and Melanie Smith, eds., *Routledge Handbook of Cultural Tourism*. London: Routledge.

Sandbrook, Dominic. 2010. *State of Emergency: The Way We Were: Britain 1970–1974*. London: Allen Lane.

Till, Jeremy. 2009. *Architecture Depends*. Cambridge, Mass: MIT Press.

Vaughan, Laura et al. (2009) 'Do the Suburbs Exist? Discovering Complexity and Specificity in Suburban Built Form'. *Transactions of of the Institute of British Geographers*, 34(4), 475–488.

VisitLondon. 2015. 'Visiting London for the First Time', 15 June. Available at http://www.visitlondon.com/discover-london/london-highlights/visiting-london-for-the-first-time?ref=nav-dl (accessed 22 June 2015).

Webster, Roger. 2000. 'Suburbia Inside Out'. In Roger Webster, ed., *Expanding Suburbia: Reviewing Suburban Narratives*. New York, NY: Berghahn Books.

Whitehand, Jeremy and Christine Carr. 2001. *Twentieth Century Suburbs, a Morphological Approach*. London: Routledge.

Willmott, Peter and Michael Young. 1960. *Family and Class in a London Suburb*. London: Routledge and Kegan Paul.

Wolfram, Gemot and Claire Burnill Maier. 2013. 'The Tactical Tourist'. In Greg Richards and Melanie Smith, eds., *Routledge Handbook of Cultural Tourism*. London: Routledge.

Wynn, Jonathan. 2010. 'City Tour Guides: Urban Alchemists at Work'. *City and Community,* 9(2), 145–163.

Acknowledgement

This chapter is based on research and ideas carried out and developed by the author with Professor Peter Newman, Dr Ilaria Pappalepore and Dr Andrew Smith, University of Westminster. It draws on material presented at and published after the 'Linking Urban and Rural Tourism: Strategies in Sustainability' conference, George Mason University, Virginia, USA 2015.

CHAPTER 3

Unplanned Expansions: Renting Private Homes to Tourists

Clare Inkson

Introduction

'It is important to recall that most things that are now attractions did not start out that way' (MacCannell 1999, 203).

Tourism within the private realm of a destination, that is, within residential homes via digital sharing economy platforms, is a relatively new phenomenon that has taken many destinations worldwide by surprise. It has created opportunities for new tourism service suppliers while at the same time causing controversy and concern amongst residents and local authorities. Converting the homes of local residents into tourist accommodation has become an extremely common phenomenon in London over the past decade. This trend is driven by a number of internal and external forces, policies and interventions that have developed or occurred since the early years of the twenty-first century. This chapter focuses on the expansion of tourism into London's private realm via digital sharing economy platforms; it explores links with concepts of 'authenticity', 'dedifferentiation', and 'regulated deregulation' and reveals the tensions

---

How to cite this book chapter:

Inkson, C. 2019. Unplanned Expansions: Renting Private Homes to Tourists. In: Smith, A. and Graham, A. (eds.) *Destination London: The Expansion of the Visitor Economy.* Pp. 37–59. London: University of Westminster Press. DOI: https://doi.org/10.16997/book35.c. License: CC-BY-NC-ND

between the social organisation and political governance of sharing economy accommodation, and between mass tourism and housing.

Defining the 'sharing economy' is not an easy task because the sector is not homogenous and definitions are still evolving (Gyimóthy and Dredge 2017). The term 'sharing economy' is used inconsistently and interchangeably with terms for other new 'economies' such as the 'collaborative economy', 'peer-to-peer economy', 'gift economy' and 'hybrid economy'. The 'sharing economy' broadly describes a sector that enables individuals to make their under-utilised assets, time and/or skills available for temporary use by others via digital platforms (Stephany 2015; Wosskow 2014). These assets may be gifted, mutually swapped with the same types of asset of the other party, bartered in exchange for other services, or offered for a fee which usually provides some profit for the asset owner and some value for the digital platform that mediated the transaction. The connection between asset owners and potential temporary users is facilitated by digital platforms that match owners and users. Some platforms offer a free service to both parties, some require paid membership and others charge a commission or fee to one, or both, of the parties. Where money is exchanged, the platform usually handles all transactions.

Tourism and the sharing economy seem to be natural partners: visitors to destinations require temporary use of transportation, accommodation, food, guiding and entertainment services. In many destinations these have traditionally been supplied by the commercial sector. However, the opportunities offered by digital platforms to connect private owners of vehicles, spare accommodation capacity in their homes, culinary or entertainment skills, or specific knowledge about the destination, with visitors from around the world have created new tourism supply capacity within destinations, and financial opportunities for asset owners. These are described in the literature as 'micro-entrepreneurs' if the motive is to profit financially from the 'share' (Stephany 2015). Many authors (e.g. Gyimóthy and Dredge 2017; Gyimóthy 2017; Stephany 2015) have debated the contradiction between the concept of 'sharing' and the profit motive; and it is not the objective of this chapter to debate this point. However, it is important to note that distinctions between the sharing economy and the commercial sector are sometimes difficult to distinguish; some sharing economy owners and platforms operate on commercial principles; some are now owned by, or in partnership with, multinational corporations; and some of the world's largest and most powerful online travel agencies (OTAs) own subsidiaries that connect asset owners in destinations with tourists seeking accommodation or other activities there.

One of the most prevalent forms of sharing economy supply is accommodation. Home owners may make a bed, a room, rooms, or their entire home, available at times when it is unoccupied, and advertise it via one or more multinational digital platforms. In London, thousands of private homes are now available for short-term rental.

The Appeal of Tourism within the Private Realm

Tourist demand for overnight accommodation in the homes of local residents is not a new phenomenon; prior to its promotion to mass markets via digital sharing economy platforms, local homes were already available to niche markets or closed groups. For example, VFR accommodation in local homes depended on personal acquaintance with or introduction to the home owner, while language schools organised stays with host families for their students. The introduction of online sharing economy platforms expanded the scale and scope of the private realm in city destinations by stimulating the growth in supply of local homes, often introducing free-market principles to the supply, and distributing it publicly on a global scale to tourist markets whose choices were previously often limited to the hotel sector – particularly in city destinations.

This expansion of tourist accommodation in the private realm coincided with a shift in tourist demand for more flexible and distinct experiences and the emergence of what are often described as 'new' types of tourists. Often rooted in discussions of post-modernism, authenticity, class and society, these new tourists reject settings, products and experiences that are designed specifically for tourists, and seek the 'real', 'genuine', authentic aspects of a destination (see also the discussion in Chapter 2).

Poon identified a long-term transformation in demand away from 'mass, standardised and rigidly inflexible' (Poon 1993, 4) forms of tourism to more independent, flexible and customised tourist experiences. Her 'new tourists', identified in the early 1990s, are experienced travellers who are more demanding; they seek control over their tourism experiences, are adventurous and open to new and different activities, and seek individual experiences as evidence of their individuality. New or innovative tourist products and opportunities to experience the 'unusual' or 'different' within a destination are likely to appeal to such market segments.

Some authors contend that tourists are ultimately seeking authenticity. MacCannell (1999) suggests that tourists seek a deep understanding of society and culture – 'it is a basic component of their motivation to travel' (1999, 100) – yet there are challenges in knowing how 'real' those experiences are. MacCannell differentiates between the 'genuine' and the 'spurious' (1999, 55), 'truth' and 'reality' (1999, 91) 'authenticity', (1999, 96) and 'inauthenticity' and 'staged authenticity' (1999, 98) in relation to tourist settings and suggests that the division between genuine and spurious is marked by the 'realm of the commercial' (MacCannell 1999, 155) – he suggests that that 'genuine' tourist experiences occur outside commercial settings. This chimes with Robert Maitland's research in London (see Chapter 2) which found that for some tourist market segments 'an important element in the appeal of the city is the opportunity to experience and feel a part of everyday life' (Maitland 2010, 1).

MacCannell (1999) suggests that the 'differentiation' between social groups that pervades 'modern' society, based on distinctions in socio-economic characteristics, sex, sexuality, life-style, race, education, age, occupation, and political affiliations, can be transcended by 'modern mass leisure tourism' (1999:12). '(T)he lure of the local' (Lippard 1997, cited in MacCannell 1999, 199) and 'someone else's local specificity' (MacCannell 1999, 199) can reduce perceptions of differences between groups, transforming the tourist's personality and relationships, their consciousness, and understanding of the world. Richards (2017) links sharing economy accommodation with the potential for tourists to achieve self-actualisation, 'where personalised experiences are generated through empathy between host and guest' (2017, 173).

Maitland (2010) found synergies between some tourists and some residents in London, particularly those seen as the 'cosmopolitan consuming class' (Fainstein et al. 2003, cited in Maitland 2010, 176) or the 'creative classes' (Florida 2002, cited in Maitland 2010, 176); they engage in similar activities in cities and share tastes and preferences. This 'de-differentiation' between tourism consumption and other forms of consumption results in 'dissolving boundaries between tourists, residents and other city users, and between touristic and non-touristic behaviours' (Maitland 2010, 178).

The transformation from 'old' to 'new' tourists is often linked to a desire for new areas/destinations that offer authenticity and a distinct sense of place. MacCannell, drawing on Goffman's division of social settings into front and back, identifies a continuum of six stages to the search for tourist settings that reveal the 'back region'; stage six provides access to 'the kind of social space that motivates touristic consciousness' (1999, 102). Maitland's research into the attraction of new areas of London revealed the growing importance of 'distinctiveness' and 'conviviality' as part of the tourist experience and the desire to escape from the 'tourist bubble' (Maitland 2008). His work revealed the emergence and growth of a new type of tourist in London: those keen to avoid tourist hot-spots within destinations by seeking 'unspoilt', 'genuine' or 'authentic' places and experiences. Research in Islington, in north London, found that for visitors there 'the presence of local people was important … and even the most mundane features of everyday life were of interest…' (Maitland and Newman 2009, 79). In Bankside, part of the South Bank area, his research with Newman found a desire from tourists to observe 'the presence of people who were not tourists, but Londoners working, shopping and relaxing … an area in which the "real" city could be experienced' (2009, 82). The private realm within a destination, that is, the lives, and lifestyles of local people might be seen as an extension of this search for authenticity, sense of place and the 'real' city. This helps to explain the popularity of sharing economy accommodation. Staying in private homes broadens the scope of access to the private realm through first-hand exposure to the 'back region' (MacCannell 1999) of a destination and the 'real' lives of residents through access to their homes. Indeed, rental of entire homes, and therefore the absence of the home owner, may create a sense of borrowing the home owner's lifestyle and participating in the local's experience of a place.

Maitland and Newman (2009) suggest that London's ability to attract a high percentage of repeat visitors puts the city in a strong position to disperse tourists outside the 'tourist bubble': 44 per cent of overseas and 85 per cent of domestic overnight tourists had stayed in London at least twice in the preceding five years, while 81 per cent of domestic day visitors had visited at least ten times (LDA 2007, cited in Maitland and Newman 2009, 72). Repeat visitors' familiarity with the city may make them more amenable to new, less explored areas of the city. Urry and Larsen (2011) suggest that the 'tourist gaze' demands difference and 'out of the ordinary'. This can be achieved through 'seeing unfamiliar aspects of what had previously been thought of as familiar' (2011, 16); by experiencing new areas of a city whose main 'touristified' (Novy 2017) areas have already been visited, echoing MacCannell's claim that 'the quest for authenticity is marked off in stages in the passage from front to back' (1999, 105). Sharing economy accommodation facilitates the rapid expansion of tourism into areas of cities which have not been 'touristified' (Novy 2017), and are not typically associated with tourism, while the experience of staying in a local home might allow visitors to avoid 'the realm of the commercial' (MacCannell 1999, 155).

Maitland and Newman found that tourism policy in London did not account for the growing appeal of the '"real" city' (2009, 82) and that often the expansion of tourism into new areas of the city occurred in spite of, rather than because of, tourism policy. However, in 2017, the appeal of the private realm was officially acknowledged by London and Partners' Tourism Vision. The vision is that visitors will be able to unlock the best version of London for them by tailoring their experience to meet their needs (London and Partners 2017, 9). The vision identifies a specific role for home owners in achieving this:

> 'Encourage the sharing economy's accommodation hosts as well as hotel concierges to act as advocates for their areas, enabling their guests to experience London "like a local". Extend the suite of curated neighbourhood itineraries for visitors'. (London and Partners 2017, 39).

The role of the private realm in London's tourism future is reinforced by the inclusion of Airbnb in the industry consultation on the vision, and in the inclusion in the document of a quotation from Airbnb:

> Home sharing provides visitors with an alternative form of accommodation, disperses tourism across the city to the outer boroughs, and financially benefits Londoners. Hosts in London are ambassadors for their city, offering visitors a more authentic and local travel experience in communities beyond the city centre. By hosts sharing their local knowledge and off-the-beaten-track hidden gems, over 4 million

guests to the capital have really felt like they've lived like a local. (James McClure, General Manager, Northern Europe, Airbnb cited in London and Partners 2017, 39).

Provision of Local Homes to Tourists

The expansion of tourism accommodation within the private realm and its transformation from niche sub-sector to mass market has been driven by digital platforms that link home owners with potential guests. Airbnb is the best-known platform but there are several other significant suppliers (see Table 3.1). These platforms tend to specialise by type of exchange offered: 'commons', 'generative' or 'communitarian' business models are motivated by altruism, solidarity or reciprocity – no money is exchanged and any surplus generated by the platform, for example through advertising or membership revenue, may be re-invested into the platform. 'Extractive' business models are motivated by profit – hosts charge for the short-term rental of their property, and the platforms extract a proportion of the value created by asset-owners and distribute it to themselves (Gyimóthy 2017). Dredge (2017, 76) describes extractive models as 'platform capitalism', an extension of neoliberalism and backed by venture capital allowing such models to expand rapidly.

The importance of these platforms in transforming the accessibility of private homes to tourists should not be underestimated. Urry (1995) stresses the significance of the 'social organisation of travel' (142) and 'organisational innovations' (142) in stimulating major transformations in tourism. Examples cited by Urry (1995) include the voucher system and inclusive tours introduced by Thomas Cook in the mid-nineteenth century, and the post-war development of inclusive tours by air that made international travel accessible to the mass markets in northern Europe. The power of sharing economy platforms rests in their reduction of the risk of staying in a stranger's home through the development of a strong brand, and the creation of trust through the publication of user reviews.

The appeal of city destinations to extractive models is acknowledged by Ferreri and Sanyal (2018) who describe short-term tourism lets as a force that 'straddles the divide between housing and hospitality' (2018, 2) while others suggest that some cities are more likely to develop sharing economy resources. 'The most significant growth of collaborative business phenomena takes place in cities and urban areas with a high concentration of resources (capital, property, skills) and year-round demand with high purchase power' (Gyimóthy and Dredge 2017, 33). Data on the supply of local homes as tourist accommodation in London suggests that communitarian and extractive supply coexist in London, although extractive business models dominate. A snapshot of some of the suppliers of tourist accommodation in local homes in London is shown in Table 3.1.

Despite the distinction in business models, the significance of 'authenticity', 'backstage' and 'conviviality' to all forms of sharing-economy accommodation is evident: the language used to promote the private realm focuses on the local and 'genuine', for example, to 'live like a local' (Airbnb); 'stay in distinctive private homes' (One Fine Stay); 'unlock the secrets of the city' through staying in 'vibrant neighbourhoods filled with personality' (Under the Doormat); or, to 'discover amazing people' (Couchsurfing). Richards (2011) links the sharing

Site	Number of Entire Properties in Greater London	Source	Business Model
Couchsurfing	146,000 +	Couchsurfing.com 2017	Gifting
Airbnb	53,080 active rentals: 27,876 entire homes; 24,549 private rooms; 655 shared rooms	Airbnb 2017 cited in AirDNA 2018	Short-term rental
Lovehomeswap	650+	Love Home Swap 2017	Mutual exchange/ points accrual or purchase
Homelink	171	Homelink 2018	Mutual exchange
Guardian Home Exchange	171	Guardian Home Exchange 2018	Mutual exchange
Homeaway	5,000+	Homeaway 2018	Short-term rental
Owners Direct	3,000+	Owners Direct 2018	Short-term rental
Holiday Lettings	6,330	Holiday Lettings 2018	Short-term rental
Housetrip	6,423	Housetrip 2018	Short-term rental
Flip Key	7,137	Flip Key 2018	Short-term rental
Niumba	2,764	Niumba 2018	Short-term rental
Booking.com	4,427	Booking.com 2018	Short-term rental
Oasis Collections	50+	Oasis Collections 2018	Short-term rental
One Fine Stay	700+	One Fine Stay 2018	Short-term rental
Under the Doormat	80+	Under the Doormat 2018	Short-term rental
Veeve	370	Veeve 2018	Short-term rental

Table 3.1: Snapshot of Suppliers of Tourist Accommodation in Local Homes in London. Source: Compiled by the author.

economy to Lengkeek's (1996) 'colonisation of the lifespace' where the private sphere is commodified, and 'the locals seem to willingly collaborate in the colonisation process' (Richards 2011, 181) through the provision of their homes to tourists via the digital platforms. This calls into question the suggestion that the private sphere for tourism exists outside of commercial settings.

To provide some context, figures for March 2016 show that there were approximately 140,000 hotel rooms across London (London and Partners 2016), while AirDNA data shows that around 50 per cent of entire active rentals on Airbnb throughout London are in properties with two or more bedrooms, meaning that the capacity of short-term rental accommodation is significantly greater than the number of properties listed, at almost 42,000 rooms.

At this stage a word of caution should be noted. While broad figures about the number of properties listed on each website are fairly accessible, occupancy data are not readily available. In addition, property listings may be duplicated across a number of sites to reach a wider market; for instance, Under the Doormat lists its properties on Booking.com, Airbnb, Expedia, TripAdvisor and Homeaway, as well as on its own website (Under the Doormat 2017a). Therefore, it is extremely difficult to estimate the number of entire properties available for short-term rental in London or the number of nights that are booked. This creates challenges for destination planners, regulators and marketers.

Unplanned Expansions

Richards (2011, 180) links the sharing economy with 'interstitial private resources', that informally fit into the gaps and spaces left by the formal tourism sector. Indeed, the supply of accommodation in local homes represents capacity that is not planned for via a destination's policy and planning framework. Unplanned tourism development is described by Barbaza (1970 cited in Pearce 1989) as 'spontaneous development' that often creates a number of unanticipated short- and long-term negative impacts within the destination concerned. Spontaneous tourism development is often associated with the rampant physical growth of resorts in response to growing demand, before suitable planning measures could be put in place, for example in the Cote D'Azur in southern France in the 1950s and the Costa Brava in Spain in the 1960s (Barbaza 1970 cited in Pearce 1989). It is not normally associated with urban tourism or with global cities. However, although short-term rental of residential property uses existing property and does not necessarily require new physical development, it does provide new and unplanned-for tourism capacity that has a number of unanticipated negative consequences.

The expansion of mass tourism into the private realm bypasses public tourism and urban planning policies which are designed to achieve specific objectives leading to the attainment of a long-term vision (Yan and Morpeth 2015) and which provide a framework of regulations, guidelines, and directives to inform decisions and activities within destinations. The provision of commercial tourist accommodation within a destination is usually the result of careful planning processes that identify desirable land use in specific zones, specify the desired capacity of tourist accommodation properties or the desirable number of rooms in an area, and the preferred type and quality of accommodation in line with local tourism, planning, economic, and housing policies. The expansion into the private realm adds unplanned capacity that potentially increases tourism capacity within a destination substantially, with no control over its location or subsequent impacts.

The spontaneous expansion of mass tourism into the private realm might be seen as part of a broader neoliberal focus on reducing barriers to growth, competitiveness and attracting investment, which Gyimóthy and Dredge (2017) attribute to government tourism policies. Dredge and Jenkins (2007) stress that policy in general reflects the choices made by governments and their collaborators to balance the interests of interested stakeholders: policy '… is inherently political' (2007, 8). In London, the expansion of tourism supply in the private realm was not explicitly planned for within tourism or urban planning policies, but was enabled by UK government policies that since 2010 have been directed by free-market ideology intended to reduce regulation, reduce government expenditure and promote the growth of the free-market economy (The Conservative Party 2010; The Conservative Party 2015). A key theme of the 2011 Tourism Policy was 'Better Regulation – Cutting Red Tape' (Department for Culture, Media and Sport [DCMS] 2011), with the stated aim of the then coalition government to remove 'unnecessary rules that make it more difficult or expensive for tourism businesses to grow' (DCMS 2015). The Tourism Action Plan 2016 refers to 'common sense regulation: examining the scope for deregulation' to facilitate the growth of tourism businesses (DCMS 2016, 3).

The supply of entire residential properties in London for short-term rental was heavily regulated until 2015. Short-term rental of an entire property for fewer than 90 days to the same tenant had been illegal since 1973 in order to protect housing stock for the benefit of London's permanent residents (The National Archive 2015). Any residential property that was offered as temporary sleeping accommodation for fewer than 90 days required planning consent for a change of use from residential to temporary sleeping accommodation. Despite this, the supply of short-term rentals had already created problems in central London boroughs. In the preceding fifteen years, City of Westminster's Planning Enforcement Team investigated 7,362 properties suspected of illegal short-term lets; over 6,000 of these reverted to lawful permanent residential use (BHA n.d.). A number of online platforms already offered entire properties in London for short term rental: in June 2014, more than 6,600 entire properties (either houses or flats) were listed on Airbnb (Ball et al. 2017); One Fine Stay had been promoting and managing entire properties in London since 2009 (One Fine Stay 2017); Veeve since 2011 (Veeve 2018); and Under the Doormat since 2014 (Under the Doormat 2017b). The ability of these companies to bypass regulations is a product of the challenge for policy makers to keep pace with technological change and the willing supply of homes by micro-entrepreneurs (Guttentag 2017).

Short-term letting in London was addressed specifically in the 2015 Deregulation Act which amended planning laws in London. The objectives of these changes were to: provide opportunities for London residents to rent out their homes; help boost London tourism through the provision of 'competitively-priced accommodation'; reduce the number of properties lying empty or underused; reform laws which are 'poorly and confusingly enforced across London'; and create similar freedoms and flexibility to the rest of the UK (Department for Communities and Local Government 2015). The revised regulations allowed for '… some common-sense measures to protect local amenity, whilst allowing Londoners who go on holiday to make a bit of extra money by renting out their home whilst they are away' (Ministry of Housing, Communities and Local Government 2015). These measures, in a process of 'regulated deregulation' (Aalbers 2016 cited in Ferreri and Sanyal, 2018, 4), reduced state control over short-term letting in London, but introduced a framework of regulations to which home owners must comply. The Deregulation Act 2015 allowed 'the use as temporary sleeping accommodation of any residential premises in Greater London' if the cumulative total of nights does not exceed ninety in each calendar year and if the provider of the accommodation pays council tax (The National Archive 2015). These regulations apply to entire properties only; hosts are able to let individual private rooms in their homes without any restrictions, although tenancy agreements in rented properties often forbid this. The relaxation of planning laws responded to 'changes in the way people want to use their homes', and the development of online platforms enabling this (The National Archive 2015).

Illustrating the 'contradictory priorities at different levels of government' (Ferreri and Sanyal 2018, 4), local authorities in London lobbied against the proposed 90-day rule, arguing that a 30-day limit would more closely reflect average annual holiday entitlement and therefore the amount of time that many properties would be unoccupied, and argued that relaxation of short-term letting regulations risked damaging communities through '… anti-social behaviour, fear of crime and loss of neighbourhood character' (London Councils 2014, 3), and the loss of permanent housing stock.

The pressure on housing in London and the need to protect housing stock is arguably greater now than in 1973; between 1997 and 2016, London experienced a 40 per cent growth in jobs and a population increase of 25 per cent but an increase in housing supply of 15 per cent (GLA 2017). This has created inflationary pressure on property values and rents. Private rents have risen five times faster than average earnings; it is estimated that in 2014/15, one third of households renting privately in London spent more than 50 per cent of their income on rent (GLA 2017). The average price of a property in London increased by 47 per cent between June 2011 and June 2015 (The Guardian 2016). In 1990, approximately 11 per cent of households rented from private landlords; in 2016 this was estimated to be 28 per cent. Home ownership in London has fallen from 57 per cent in 2001 to 50 per cent in 2011 and is expected to fall below 40 per cent by 2025, with significant reduction in the number of owner occupiers under the age of 34 (GLA 2017). While local authorities support opportunities for home owners to earn income from their homes legally, seven London boroughs (Camden, Hammersmith and Fulham, Haringey, Islington, Lewisham, Waltham Forest and City of Westminster) have requested tougher legislation to discourage illegal lettings (BBC News 2017). Four of the boroughs listed above are in the outer areas of London, suggesting that they are anticipating, or have already experienced, an expansion of short-term lets.

National government recommended that in cases of 'egregious breaches of regulation … the government, local authorities and sharing economy platforms should work together to ensure that all legal requirements are met' (Wosskow 2014, 10). Enforcement of the regulations is the responsibility of the local authority for the area concerned but the quote above highlights the role of digital sharing economy platforms in governance of the sector and monitoring compliance with regulations, 'allowing (them) to actively intervene in the very definition of regulation' (Ferreri and Sanyal 2018, 13). The willingness of short-term rental providers to monitor the occupancy of their property listings in London is mixed. In 2016, Airbnb – under pressure from local authorities to prevent illegal lettings – introduced a maximum of 90 days' availability per calendar year for each London property on its site that did not have planning consent for change of use, but this does not prevent a property owner from listing on other sites. BBC London found that Airbnb's competitors had no plans to enforce the 90-day rule (cited in Lynn and Allen 2017). Homeaway provides information on its website advising home owners of the 90-day rule and other regulations relating to short-term lets in London. Trip Advisor Rentals requires homeowners to agree to comply with local laws and regulations (Lynn and Allen 2017). But these sites do not actively monitor availability and bookings.

Property owners avoid scrutiny by listing a single property on several platforms, or re-registering it on Airbnb using different names and descriptions (Lynn and Allen 2017). Inside Airbnb data (cited by Ferreri and Sanyal 2018) showed that in 2015 and 2016, 41.3 per cent of the entire properties in London listed on Airbnb were multiple listings by single hosts. In 2016, the Institute for Public Policy Research found that almost 25 per cent of short-term rental properties breached the 90-day limit (They Work for You 2017). In September 2017, the City of Westminster suspected that almost 1,500 properties breached the 90-day rule (City of Westminster 2017).

Enforcing of the 90-day rule by planning authorities is slow and complex, often relying on notifications from the public, and requiring investment in staff resources to investigate alleged breaches, monitor properties across several online platforms, and visiting properties suspected of breaching the regulations. The City of Westminster employs six full-time planning enforcement officers who work exclusively on short-term let investigations within the borough (Ferreri and Sanyal 2018). Camden Council has created a 'partial database' based on data from Inside Airbnb to collect evidence to support enforcement and prosecution (Ferreri and Sanyal 2018). The enforcement process itself takes several months. The earliest that a breach of the permitted 90 days can be identified is 1 April, based on 100 per cent occupancy, and further time is required to serve an enforcement notice and process any appeals. The calendar year is almost over by the time a conviction can occur, and the process starts again the following year.

Spatial Distribution of Short-term Rentals of Local Homes in London

Digital sharing economy platforms have a significant role in 'making spaces' (Ferreri and Sanyal 2018, 3) and in expanding the spatial distribution of tourism accommodation capacity further across London to accelerate the emergence of Maitland's 'new areas' for tourism. Table 3.2 shows that while central areas feature heavily in short-term letting provision, new tourist accommodation areas of the city are emerging. To some extent this is influenced by short-term letting companies that offer managed or serviced properties and therefore develop a cluster of properties to create a geographical critical mass that achieves efficiencies in the delivery of the service – a further example of 'organisational innovation' (Urry, 1995). For example, Under the Doormat offers around 30 properties each in the boroughs of Wandsworth, Lambeth and Richmond-upon-Thames in south-west London and occupancy data for 2016 published by My Property Host (cited in Lewis 2016) revealed that their most popular boroughs by occupancy rate are Tower Hamlets and Hackney in the east of the city, followed by Kensington and Chelsea, the City of London, and Islington. As the product becomes more popular for business tourists and families, they expect outer boroughs with larger properties to become more significant (My Property Host 2016 cited in Lewis 2016).

Each London Borough has its own elected council and is responsible for planning, housing and other services. Quattrone et al. (2016) conducted longitudinal research into the spatial distribution of Airbnb in London between 2012 and 2015. They found that the penetration of Airbnb listings in different areas of London was influenced by proximity to the centre, accessibility by public transport, the socio-economic profile of the area's residents, the number of attractions, the proportion of rental properties compared to owned, and the youth and tech-savviness of its residents. They identified a number of stages in the evolution of Airbnb property listings in London over that time. In the early stages, geography was the most significant determinant with a concentration of listings in areas close to the centre. These areas contained a high proportion of young and ethnically-diverse residents who were technologically literate

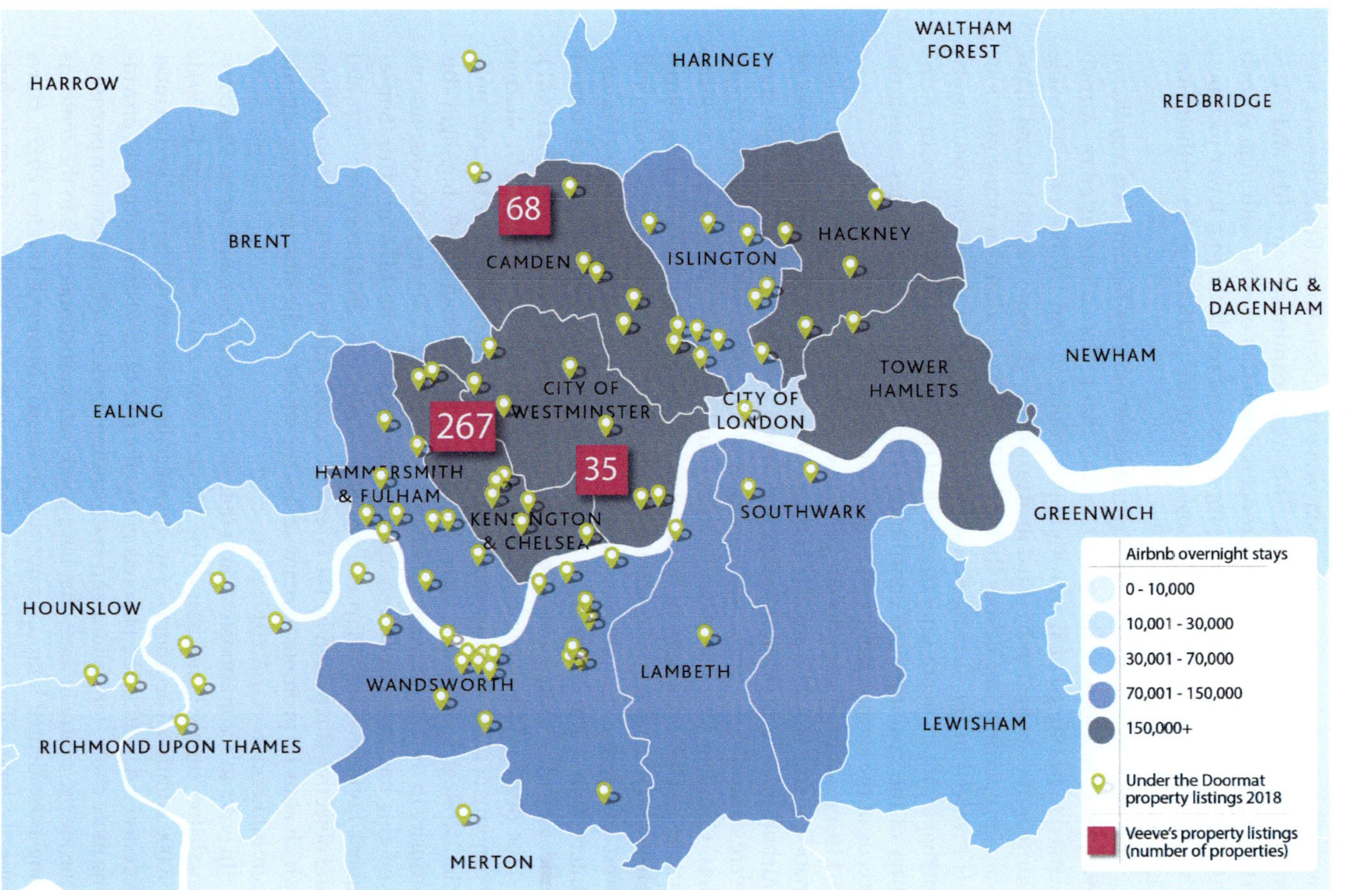

Figure 3.1: The Spatial Distribution of London's Sharing Economy Accommodation Sector (© Mason Edwards).

and early-adopters of hosting. In 2013, Airbnb began to penetrate areas further away from the centre, with less tech-savvy residents, more home owners, and more residents on lower incomes. This trend continued into 2014 and 2015. Their findings suggest that the strongest predictors of areas with high Airbnb listings are the number of rental properties and residents with lower incomes; this suggests that late-adopting hosts joined the platform to boost their own incomes.

Proximity to the centre and technological prowess became less significant over time. In addition, the authors found that private rooms listed on Airbnb tended to be offered in areas with highly-educated non-UK born renters while entire homes tended to be listed in areas with high home ownership and high-value properties. This echoes Richards' (2017) suggestion that many individuals involved in the sharing economy are members of the ex-pat community, perhaps with in-depth understanding of the needs of source markets, or members of the 'cosmopolitan consuming class', perhaps whose homes match the needs of the similar groups visiting as tourists.

Listings of properties do not provide evidence of occupancy. Colliers International and Hotelschool The Hague (2016) mapped the occupancy rates of Airbnb in each London borough and found that over half of overnight stays were concentrated in the five boroughs of City of Westminster, Tower Hamlets, Camden, Kensington and Chelsea and Hackney – adding more tourist accommodation capacity in these boroughs and strengthening existing tourist enclaves (Richards 2017). However, outer boroughs to the north, south and west of London also achieved over 30,000 overnight stays in 2015 (Colliers International and Hotelschool The Hague 2016), 'pioneer(ing) new tourist nodes which are more integrated into local communities' (Richards, 2011, 181).

The dispersal away from central London that is being facilitated by sharing economy accommodation supply has long been an objective of tourism policies, strategies and action plans. Maitland and Newman (2009) highlight this priority in the London Development Agency's 2002, 2004 and 2006 documents and this has continued even after the abolition of the LDA in 2010. Subsequent London Plans (Mayor of London 2004; 2008; 2011) mention this, and the objective is repeated by the Greater London Authority's 2017 London Plan which aims to 'promote tourism across the whole of the city, including outside central London' (GLA 2017, 262).

In order to accommodate visitors outside central London, an increase in accommodation supply is necessary. The London Plan 2017 estimates that, by 2041, 58,000 additional bedrooms of serviced accommodation will be required to meet forecasted demand. The plan recognises the role of short-term rentals in providing more choice for visitors, and in expanding serviced accommodation provision outside the centre in areas that have good transport connections to the centre as long as residential housing is not compromised (Mayor of London 2017). However, the long enforcement process and the absence of reliable data create challenges in protecting housing stock and in enforcing planning regulations.

Accommodation in the Private Realm Becomes Mainstream

Guttentag (2015) locates the sharing economy in the context of Christensen's disruptive innovation theory (1995 cited in Guttentag 2015, 1194). The theory describes the process by which a new type of product eventually transforms the market by disrupting existing suppliers and becoming mainstream. At first the disruptive product 'will generally underperform with regards to the prevailing products' key performance attributes' (Guttentag 2015, 1194) and instead offers other benefits including lower prices or convenience. Consequently, the market for the disruptive product is small initially with limited profitability and is ignored by leading companies. However, improvements to the disruptive product increase its appeal – and therefore its market share and profitability – and it becomes mainstream and a competitive threat to leading companies. Short-term rentals and continuing innovation in that sector seem to be demonstrating the disruptive innovation model in tourism.

A recurring theme in the literature is the economic shift afforded by sharing economy platforms away from 'the traditional tourism system' (Richards 2017, 174), the ability of tourists to 'circumvent the tourism supply chain' (2017, 169) and for homeowners to 'effortlessly enter the tourism accommodation sector and compete with traditional accommodation enterprises for guests from around the world' (Novy 2015, 1195). In fact, recent moves by the 'traditional' tourism supply chain to incorporate sharing economy accommodation and the private sphere into its own offer demonstrate a response to the competitive threat of sharing economy platforms. Several of the brands listed in Table 3.1 are owned by multinational OTAs that dominate the distribution of tourism accommodation and other tourist services within major markets globally: Homeaway and Owners Direct are part of Expedia, Inc; Holiday Lettings, Housetrip, Niumba and Flipkey are owned by Trip Advisor LLC; and Booking.com is part of Booking Holdings Inc. These companies are not sharing economy businesses but mainstream distribution channels of tourist accommodation.

Similarly, some global hotel companies are responding to the tourist appeal of local homes by expanding their accommodation offer and tapping into the growing demand for short-term home rental. Onefinestay.com, established in London in 2009 and offering more than 700 entire London homes, was acquired by one of the largest hotel companies in the world, Accor Hotels, in 2016 (One Fine Stay 2017). Oasis Collections, with over fifty properties in London, now operates under Hyatt's 'soft brand' The Unbound Collection (Oasis Collections 2018).

Companies that traditionally existed outside of the habitual tourism supply chain have also entered the sharing economy arena, representing 'the encroachment of professional letting into the short-term let platform economy' (Ferreri and Sanyal 2018, 9). Many of the properties listed on sharing economy and OTA platforms are managed by third party companies who handle promotion, reservations, cleaning and key handling. In such cases, home owner and guest do not need to interact at all. Some offer interior design services and further 'stage' the property through the provision of hotel-standard linen, toiletries and a welcome service. The appeal of these services to home owners in London has given rise to professional services offering property management and hospitality skills: for example Airsorted, Air Agents, Hostmaker, My Property Host, and CityRelay and Lavanda manage short-term lets on behalf of hosts by organising property listing, revenue management, communication with guests, housekeeping and keys and offer ' … an end-to-end solution, ensuring you get maximum value from your property without having to lift a finger' (Lavanda 2018). Private sector landlords, property specialists and estate agents, and property speculators have also recognised the potential of short-term lets to tourists: some London estate agencies have expanded into short-term rentals by offering management services to private-sector landlords (Kinleigh, Folkard and Hayward, 2018) while Lavanda also works with estate agents to maximise income from vacant properties awaiting sale or rental (Lavanda 2018). Private sector landlords' letting strategies have been transformed by sharing economy opportunities where higher yielding short-term rentals are substituting long-term rentals to local residents (Simcock 2017). Valuation Office data (cited in They Work For You 2017) shows that short-term rentals commonly yield weekly rates three times higher than long-term rentals in the same property: Ferreri and Sanyal (2018) found a difference per day for a one-bedroom flat of between £49 to £150 in Islington and £67 to £178 per day in Kensington and Chelsea.

Integration of the accommodation sharing economy sector into the mainstream has been reinforced by the establishment in 2017 in the UK of a trade association – the Short-Term Accommodation Association. Its role is to represent the interests of the short-term lettings sector and to work with 'all stakeholders towards a stable and supportive regulatory environment' (STAA n.d.a), further echoing Dredge's notion of 'regulatory capture' (2017, 83). The STAA Code of Conduct includes four key principles: protecting hosts and guests, supporting enforcement, maintaining residential amenity, and supporting local business. The code pledges that members will 'always promote responsible hosting and compliance with local regulations' (STAA n.d.b). While STAA requires members to remove listings that breach regulations, their code specifies such decisions depend on local authority notification. The missing link in this process is local authority access to easily obtainable evidence.

Conclusions

From a tourism perspective, the growth in supply of short-term rentals in London is reminiscent of spontaneous tourism development that has previously caused a range of long-term problems in some coastal destinations of southern

Europe. Unplanned accommodation capacity circumvents existing urban and tourism planning policies. The lack of transparency about capacity, availability and occupancy of short-term rentals prevents local authorities from managing enforcement of illegal lettings, and limits the effectiveness of destination management and marketing. The political will at a national level to embrace business, entrepreneurship, and innovation has enabled 'disruptors' to accelerate their expansion into London, while the process of 'regulated deregulation' (Ferreri and Sanyal 2018) has not provided the resources for local authorities to enforce rules and seems to have facilitated the transformation of local homes into tourist accommodation. This is no longer a niche sector serving more discerning clients, but a mass market, and one that is affecting the availability and affordability of residential housing in London.

De-differentiation is a common theme in London's experience of the sharing economy accommodation sector. Distinctions are being eroded on a number of levels: between the everyday and the 'tourist bubble', between 'staged' and 'authentic' experiences, between the private realm and global businesses, and between tourism and residential property sectors. In 'touristified' areas in the centre, existing accommodation supply, mainly in hotels, is supplemented by thousands of residential properties, reinforcing the centre as a tourist enclave, while clusters of serviced homes are facilitating the expansion of tourism into 'back regions' of the city, supported by tourism policy and potentially creating new 'tourist bubbles'.

Accommodation which is promoted as offering an 'authentic' and local experience may be professionally managed and 'staged' to more closely meet the needs of the 'cosmopolitan consuming classes' tourist market it serves – removing any personal contact between the home owner and guest and eroding the distinction between the commercial accommodation sector and the sharing economy.

The response by the 'traditional' tourism sector to the growth of the sharing economy has been to incorporate short-term letting into its own offer. The effect has been to enable the global distribution of local homes in London by massive OTAs or by global hotel companies. This is driving the growth and territorial expansion of the tourism industry in London.

References

Airdna. 2018. *Market Overview London*. Available at https://www.airdna.co/market-data/app/gb/london/london/overview (accessed 4 January 2018).

Airsorted. 2018. *Hassle free Airbnb*. Available at https://www.airsorted.uk/.

Ball, James, George Arnett, and Will Franklin. 2014. 'London's Buy-to-let Landlords Look to Move in on Spare Room Website Airbnb'. *The Guardian*, 20 June. Available at https://www.theguardian.com/technology/2014/jun/20/buy-to-let-landlords-leasing-properties-airbnb-uk.

Bauwens, Michel. 2005. *1000 Days of Theory – The Political Economy of Peer Production*. Available at www.ctheory.net/articles.aspx?id=499.

BHA. n.d. *Short Term Lettings*. Available at http://www.bha.org.uk/wordpress/wp-content/uploads/2014/07/Short-Term-Lettings.pdf

Booking.com. 2018. *London*. Available at https://www.booking.com/search results.en-gb.html?label=gen173nr-1FCAEoggJCAlhYSDNYBGh-QiAEBmAEuuAEHyAEM2AEB6AEB-AELkgIBeagCAw&sid=d6 b9002f82ec7a74857cd174a9b52f44&class_interval=1&dest_id=-2601889&dest_type=city&dtdisc=0&from_sf=1&group_adults=2&group_children=0&inac=0&index_postcard=0&label_click=undef&nflt=ht_id%3D201%3B&no_rooms=1&offset=0&pop_filter_id=ht_id-201&pop_filter_pos=4&pop_filter_rank=10&postcard=0&raw_dest_type=city&room1=A%2CA&sb_price_type=total&search_selected=1&src=index&src_elem=sb&ss=London%2C%20 Greater%20London%2C%20United%20Kingdom&ss_all=0&ss_raw=Lon&ssb=empty&sshis=0 (accessed 14 January 2018).

Cabinet Office. n.d.. *About Red Tape Challenge*. Available at http://webarchive.nationalarchives.gov.uk/20150319091553/http://www.redtapechallenge.cabinetoffice.gov.uk/about/.

City of Westminster. 2013. *Westminster's City Plan: Strategic Policies*. London: City of Westminster.

City of Westminster. 2017. *Task force to catch out landlords breaking 90 day rule*, 12 December. Available at https://www.westminster.gov.uk/task-force-catch-out-landlords-breaking-90-night-rule.

City Relay. 2018. *Who are we*? Available at https://cityrelay.com/about-city-relay/.

Colliers International and Hotelschool The Hague. 2016. *Airbnb – Impact and Outlook for London*. Available at https://www.researchgate.net/publication/310258717_Airbnb_Impact_and_Outlook_for_London.

Conservative Party, The. 2010. *The Conservative Manifesto 2010*. Available from https://www.conservatives.com/~/media/Files/Manifesto2010.

Conservative Party, The. 2015. *Manifesto 2015*. Available from https://www.conservatives.com/manifesto2015.

Couchsurfing.com. 2017. *London*. Available at https://www.couchsurfing.com/places/europe/england/london (accessed 2 November 2017).

Department for Communities and Local Government. 2015. *Measures to boost sharing economy in London*, 9 February. Available at https://www.gov.uk/government/news/measures-to-boost-sharing-economy-in-london.

Department for Communities and Local Government. 2017. *Ensuring Effective Enforcement*. Available at https://www.gov.uk/guidance/ensuring-effective-enforcement#Planning-Enforcement-Order (accessed 7 December 2017).

Department for Culture, Media and Sport. 2011. *Government Tourism Policy March 2011*. Available at https://www.gov.uk/government/uploads/system/

uploads/attachment_data/file/78416/Government2_Tourism_Policy_
2011.pdf.

Department for Culture, Media and Sport. 2015. *Policy Paper 2010-2015 Govern-ment Policy: Tourism:* https://www.gov.uk/government/publications/2010-to-2015-government-policy-tourism/2010-to-2015-government-policy-tourism

Department for Culture, Media and Sport. 2016. *Tourism Action Plan August 2016.* Available at https://www.gov.uk/government/uploads/system/uploads/attachment_data/file/555039/Tourism_Action_Plan.pdf.

Department for Digital, Culture, Media and Sport. 2017 *Tourism Action Plan One Year On October 2017.* Available at https://www.gov.uk/government/uploads/system/uploads/attachment_data/file/653396/Tourism_Action_Plan_-_One_Year_On.pdf.

Dredge, Dianne and Szilvia Gyimothy, eds. 2017. *Collaborative Economy and Tourism – Perspectives, Politics, Policies and Prospects.* Cham: Sprinter International Publishing.

Dredge, Diane and John Jenkins. 2007. *Tourism Planning and Policy.* Milton, Aus.: Wiley.

Dredge, Diane. 2017. 'Policy and Regulatory Challenges in the Tourism Collaborative Economy'. In Dianne Dredge and Szilvia Gyimóthy, eds., *Collaborative Economy and Tourism – Perspectives, Politics, Policies and Prospects.* Cham: Sprinter International Publishing.

Ferreri, Mara and Romola Sanyal. 2018. 'Platform Economies and Urban Planning: Airbnb and Regulated Deregulation in London'. *Urban Studies,* 55(15), 3353–3368.

Flipkey. 2018. *London Rentals.* Available at https://www.flipkey.com/ (accessed 19 January 2018).

GLA. 2017. *London Draft Housing Strategy.* Available at https://www.london.gov.uk/sites/default/files/2017_london_draft_housing_strategy.pdf.

Guardian, The. 2016. *London Population Growth Rate twice that of UK, Official Figures show,* 12 November. Available at https://www.theguardian.com/uk-news/2016/oct/12/london-population-growth-twice-that-of-uk-official-figures-show.

Guardian Home Exchange. 2018. *Home Swap Search.* Available at https://www.guardianhomeexchange.co.uk/search?continent=6&country=England®ion=London&lastMinOffer=No&openToOffer=Yes&item_per_page=9&view=gridview&search_option=desc&search_by=location&pageon=20 (accessed 4 January 2018).

Guttentag, Daniel. 2015. 'Airbnb: Disruptive Innovation and the Rise of an Informal Tourism Accommodation Sector'. *Current Issues in Tourism,* 18(12), 1192–1217.

Gyimóthy, Szilvia and Dianne Dredge. 2017. 'Definitions and Mapping the Landscape in the Collaborative Economy'. In Dianne Dredge and Szilvia Gyimóthy, eds, *Collaborative Economy and Tourism – Perspectives, Politics, Policies and Prospects.* Cham: Sprinter International Publishing.

Gyimóthy, Szilvia. 2017. 'Business Models in the Collaborative Economy'. In Dianne Dredge and Szilvia Gyimóthy, eds, *Collaborative Economy and Tourism – Perspectives, Politics, Policies and Prospects*. Cham: Sprinter International Publishing.

Holiday Lettings. 2018. *Apartments in London and Flats – Holiday Lettings – Holiday Rentals London*. Available from https://www.holidaylettings.co.uk/london/.

Homeaway. 2017. *Rules for London Owners*. Available at https://www.homeaway.co.uk/info/about-us/legal/owner-responsibilities/rules-for-london-owners (accessed 7 December 2017).

Homeaway. 2018. *London UK*. Available at https://www.homeaway.co.uk/results/united-kingdom/london/region:482/minSleeps/4/page:100 (accessed 14 January 2018).

Homelink.org. 2018. *London*. Available at https://homelink.org/en/search/mapview (accessed 4 January 2018).

Housetrip. 2018. Appartements de Vacances à Londres. Available at https://www.housetrip.fr/ (accessed 19 January 2018).

Kinleigh, Folkard and Hayward. 2018. *Completely London*. Available at https://www.kfh.co.uk/kfh-services/lettings/short-term-lets (accessed 14 January 2018).

Kollewe, Julia. 2016. 'UK Interest Rate Moves since 2007 – Timeline'. *The Guardian*, 4 August Available at https://www.theguardian.com/business/2016/jul/14/uk-interest-rates-timeline-bank-of-england-uk-economy.

Lavanda. 2018. *Our Products*. Available at https://lavanda.co.uk/?utm_source=Google&utm_medium=PPC%20Ads&utm_campaign=Lavanda%20Brand&campaign=755436226&content=207176232892&keyword=lavanda&adposition=1t1&matchtype=e&device=c&devicemodel=&interest=&location=9046003&gclid=EAIaIQobChMI396M7dq-2AIVp7ftCh1N-qQP7EAAYASAAEgKC5vD_BwE.

Leegkeek, Jaap. 1996. 'On the Multiple Realities of Leisure; A Phenomenological Approach to the Otherness of Leisure'. *Loisir et Société/Society and Leisure* 19(1), 23–40.

Lewis, Warren. 2016. 'Who are the high risers in London's Airbnb league table?' In *Property Reporter*, 26 October. Available at http://www.propertyreporter.co.uk/landlords/who-are-the-high-risers-in-londons-airbnb-league-t4ble.html

London Councils. 2014. *Shortterm Lets in London: A London Councils Member Briefing* September. Available at https://www.londoncouncils.gov.uk/node/1822.

London and Partners. 2016. *London Hotel Development Monitor – Uncovering Hotel Opportunities*. Available at https://files.londonandpartners.com/l-and-p/assets/hotel_development_monitor_may_2016.pdf

London and Partners. 2017. *A Tourism Vision for London*. Available at http://files.londonandpartners.com/l-and-p/assets/london_tourism_vision_aug_2017.pdf.

Love Home Swap.com. 2017. *Greater London*. Available at https://www. lovehomeswap.com/homes/Greater-London--England--United-Kingdom? guests=2&within=1&page=1 (accessed 2 November 2017).

Lynn, Guy and Aurelia Allen. 2017. 'Airbnb Time Limits "Ineffective in London" Councils say'. *BBC News*, 10 February: https://www.bbc.co.uk/news/uk-england-london-38924720

MacCannell, Dean. 1999. *The Tourist – A New Theory of the Leisure Class*. Berkeley, CA: University of California Press.

Maitland, Robert and Peter Newman. 2009. 'London – Tourism Moving East?' In Robert Maitland and Peter Newman, eds, *World Tourism Cities – Developing Tourism off the Beaten Track*. Abingdon: Routledge.

Maitland, Robert and Peter Newman, eds. 2009. *World Tourism Cities – Developing Tourism off the Beaten Track*. Abingdon: Routledge.

Maitland, Robert. 2008. 'Conviviality and Everyday Life: The Appeal of New Areas of London for Visitors'. *International Journal of Tourism Research*, 10(1), 15–25.

Maitland, Robert. 2010. 'Everyday Life as a Creative Experience in Cities'. *International Journal of Culture, Tourism and Hospitality Research*, 4(3), 176–185.

Mayor of London. 2004. *The London Plan Spatial Development Strategy for Greater London*, February. London: Greater London Authority.

Mayor of London. 2008. *The London Plan Spatial Development Strategy for Greater London Consolidated with Alterations since 2004*, February. London: Greater London Authority.

Mayor of London. 2011. *The London Plan Spatial Development Strategy for Greater London*, July. London: Greater London Authority.

Mayor of London. 2017. *The London Plan – The Spatial Development Strategy for Greater London*, Draft for Public Consultation, December. Available at https://www.london.gov.uk/sites/default/files/new_london_plan_december_2017_web_version.pdf.

Ministry of Housing, Communities and Local Government. 2015. *Measures to Boost Sharing Economy in London*. Available at https://www.gov.uk/government/news/measures-to-boost-sharing-economy-in-london (accessed 9 February 2018).

Morpeth, Nigel and Hongliang Yan, eds. 2015. *Planning for Tourism – Towards a Sustainable Future*, Wallingford: CABI.

National Archives, The. 2015. *Deregulation Act 2015, Explanatory Notes, Commentary on Sections, Sections 44/45*. Available at http://www.legislation.gov.uk/ukpga/2015/20/notes/division/5/46.

Niumba. 2018. *Alquileres Vacacionales Londres*. Available from https://www.niumba.com/reino-unido/inglaterra/gran-londres (accessed 19 January 2018).

Novy, Johannes. 2017. '"Destination" Berlin Revisited. From (New) Tourism Towards a Pentagon of Mobility and Place Consumption'. *Tourism Geographies*, 20(3), 418–442.

Oasis Collections. 2018. *Handpicked Homes in London.* Available at https://www.oasiscollections.com/london/search?sort=bedroomCount-DESC (accessed 4 January 2018).

One Fine Stay. 2017. *About Us.* Available at https://web.onefinestay.com/about/.

One Fine Stay. 2018. *London.* Available at https://www.onefinestay.com/search/?locations%5B0%5D=500&locationName=London&after=780 (accessed 14 January 2018).

Owners Direct. 2018. *London Bridge, London, England, United Kingdom.* Available at https://www.ownersdirect.co.uk/results/refined/keywords:London%20Bridge,%20London,%20England,%20United%20Kingdom/page:62 (accessed 14 January 2018).

Pearce, Douglas. 1989. *Tourist Development.* Harlow: Longman Scientific and Technical.

Peters, Deike. 2017. 'Density Wars in Silicon Beach: The Struggle to Mix New Spaces for Toil, Stay and Play in Santa Monica, California'. In Claire Colomb and Johannes Novy, eds, *Protest and Resistance in the Tourist City.* Abingdon: Routledge.

Poon, Auliana. 1993. *Tourism, Technology and Competitive Strategies.* Wallingford: CABI.

Quattrone, Giovanni et al. 2016. 'Who Benefits from the "Sharing" Economy of Airbnb?'. Proceedings of the 26th International ACM Conference on World Wide Web (WWW) 2016. *Cornell University Library,* arXiv:1602.02238v1.

Richards, Greg. 2011. 'Creativity and Tourism: The State of the Art'. *Annals of Tourism Research,* 38(4), 1225–1253.

Richards, Greg. 2017. 'Sharing the New Localities of Tourism'. In Dianne Dredge and Szilvia Gyimóthy, eds, *Collaborative Economy and Tourism – Perspectives, Politics, Policies and Prospects.* Cham: Sprinter International Publishing.

Sharing Economy UK. 2015. *About Us.* Available at http://www.sharingeconomyuk.com/.

Simcock, Tom. 2016. *Landlord Investment, Finance and Tax Report 2016.* Manchester: Residential Landlords Association.

Simcock, Tom. 2017. 'From Long-Term Lets to Short-Term Lets: Is Airbnb becoming the new Buy-to-Let'. Residential Landlord's Association. Available at: https://research.rla.org.uk/report/long-term-lets-short-term-lets-airbnb-new-buy-to-let/ (assessed 24 August 2017).

STAA. n.d.a. *Home.* Available from https://www.ukstaa.org/#welcome.

STAA. n.d.b. *Code of Conduct.* Available from https://www.ukstaa.org/code-of-conduct/.

Stephany, Alex. 2015. *The Business of Sharing: Making it in the New Sharing Economy.* Basingstoke: Palgrave Macmillan.

They Work for You. 2017. *Short and Holiday-Let Accommodation (Notification of Local Authorities).* Available at https://www.theyworkforyou.com/debates/?id=2017-12-13b.401.0&p=10075 (accessed 13 December 2017).

Under the Doormat. 2017a. *London United Kingdom.* Available at https://under
thedoormat.com/plan-your-stay/?gmw_address%5B0%5D=london%2C%20
London%2C%20United%20Kingdom&cf_checkin&cf_checkout&cf_
PeopleCapacity=-1&gmw_form=2&gmw_per_page=100&gmw_
lat&gmw_lng&gmw_px=pt&action=gmw_post¤cy=gbp&cf__
price%5B0%5D&cf__price%5B1%5D&cf_Bedrooms=-1&cf_TotalBath=-1
(accessed 2 November 2017).

Under the Doormat. 2017b. *Let My Home.* Available at http://underthedoormat.
com/let-my-home/.

Under the Doormat. 2018. *London, London, United Kingdom.* Available at
http://underthedoormat.com/plan-your-stay/?gmw_address%
5B0%5D=London%2C%20London%2C%20United%20Kingdom&cf_
checkin&cf_checkout&cf_PeopleCapacity=-1&gmw_form=2&gmw_
per_page=100&gmw_lat&gmw_lng&gmw_px=pt&action=gmw_
post¤cy=gbp&cf__price%5B0%5D&cf__price%5B1%5D&cf_
Bedrooms=-1&cf_TotalBath=-1 (accessed 14 January 2018).

Urry, John and Larsen, Jonas. 2011. *The Tourist Gaze 3.0,* London: Sage.

Veeve. 2018. *London.* Available at https://www.veeve.com/properties/london?
page=1&numadult=1&latitude=51.5073509&longitude=-0.127758
29999998223.

Visit Britain. 2013. *Foresight* Issue 118, August. Available at https://www.vis-
itbritain.org/sites/default/files/vb-corporate/Documents-Library/docu-
ments/Foresight-issue_118.pdf.

Visit Britain. 2017. *Modern Industrial Strategy. Tourism Sector Deal informing
the Long-Term Tourism Strategy for Britain October 2017.* Available at
https://www.visitbritain.org/sites/default/files/vb-corporate/Documents-
Library/documents/industrial_strategy_-_sector_deal_bid_submission.
pdf.

Visit England. 2014. *Self-catering Domestic Tourism.* Available at https://
www.visitbritain.org/sites/default/files/vb-corporate/Documents-Library/
documents/England-documents/self_catering_accommodation_2014.pdf.

Wosskow, Debbie. 2014. *Unlocking the Sharing Economy – An Independent
Review.* Available at https://www.gov.uk/government/uploads/system/
uploads/attachment_data/file/378291/bis-14-1227-unlocking-the-sharing-
economy-an-independent-review.pdf.

Yan, Hongliang., and Nigel Morpeth. 2015. Approaches to Planning and
Tourism in Planning for Tourism – Towards a Sustainable Future. Chapter 1
Wallingford: CABI

All internet sites without an access date last accessed on 31 March 2018

CHAPTER 4

Aerotropolis: London's Airports as Experiences and Destinations

Anne Graham

Introduction

This chapter considers the role of London's airports as experiences and destinations. While arriving by air can offer visitors an ultimate vertical city tourism opportunity (see Chapter 6), airports themselves provide the first and last experience that many tourists will encounter when they travel to a city, and so are very significant destination spaces. However, it is not always clear how visitors perceive their airport journey in relation to their overall trip. It may be that they view this journey as just a necessary and functional activity that precludes the start of their actual visit. Alternatively, the journey may be considered as an integral and non-separated part of the visit, with a positive airport experience enhancing the visitor's overall perception of the destination, and with a negative experience having the opposite effect. Arguably, the airport journey could even have some impact on the visitor's willingness to return.

The airport journey offers a range of services and facilities that are available within the terminal, with modern-day airports providing much more than

---

How to cite this book chapter:
Graham, A. 2019. Aerotropolis: London's Airports as Experiences and Destinations. In: Smith, A. and Graham, A. (eds.) *Destination London: The Expansion of the Visitor Economy.* Pp. 61–89. London: University of Westminster Press. DOI: https://doi.org/10.16997/book35.d. License: CC-BY-NC-ND

just the basic infrastructure to allow travellers to transfer from surface to air modes of transport. Some of these services, for example those related to retail and entertainment, have the potential to enhance the passenger experience and increase the airport's attractiveness, while at the same time supporting the financial well-being of the airport. Airports can also be viewed as destinations in their own right. This may be related to the deliberate development of a so-called Airport City or Aerotropolis, when airports expand beyond the boundaries of the traditional aeronautical business, by using neighbouring land for a number of additional activities, including event, conference or leisure facilities that may attract visitors. Alternatively, such development may occur in a more unintentional, organic and piecemeal fashion, encouraged by connectivity to international markets coupled with good local access that airports can provide.

For London, tourism was originally driven by the railways and grand stations that were built in the nineteenth century to impress arriving visitors, but it is now primarily airports that provide London's gateways. The situation is complex, as London offers not just one airport or gateway but a range of different ones, each having unique features and characteristics, and appealing to different airlines and passengers. The research literature has rarely considered these issues in relation to London as a destination, and so the aim of this chapter is to begin to fill this gap. It starts by exploring the general concepts that are relevant when considering an airport as an experience or destination space. This is followed by an overview of the airport system serving the London area. Bringing these two discussions together enables an assessment of the role and nature of London's airports as experiences and destinations.

General Concepts

Airports, in common with motorways, hotel rooms and shopping rooms have been defined as 'non-places' by the French anthropologist Marc Augé, being considered as spaces of transience, where people remain anonymous, and that are not significant enough to be regarded as a 'place' (Augé 1995). However, this detached placelessness notion for modern-day airports has been increasingly challenged (Appold and Kasanda 2011; Losekoot 2015), particularly with respect to whether waiting at airports can be pleasurable (Lloyd 2003) and whether airports can be viewed as so-called 'third spaces'. These spaces are separate from the two usual social environments, namely the first place or home and the second place or work, and are neutral ground where people can gather and foster community interaction, for example in cafes or bars, or in public places such as parks and libraries (Oldenburg 1989). Honegger (2013) stated that 'The airport has become, both literally and figuratively, a "third place" – a neutral crossroads of culture and function …' rather than '… non-places or a necessary pause between where one is and where one is headed' with Gottdiener (2001) arguing that airports have now become specific places in their own right with roles as a transition space and gateway; shopping mall; and city and community.

Views related to the airport space need to take into account the impact on the passenger experience, but this has received scant coverage in the literature, in spite of connected concepts, such as the visitor or tourist experience, being topics of popular debate. A notable exception is the research of Huang et al. (2018) who explored the liminal nature of airports and presented a phenomenology of passenger experience in accordance to their familiarity with the space. Within the literature on airports, the terms service quality, passenger satisfaction and passenger experience are used interchangeably, with just vague, if any, distinctions being made. Whilst these are linked concepts, their focus varies and they involve different viewpoints.

Service quality assessment relates to whether certain service standards, defined by airport management, have been achieved. Passenger satisfaction is clearly influenced by the service delivered, but it is also affected by passenger expectations. This is the basic thinking behind popular generic models such as SERVQUAL that have been used in the airport context (e.g. Pabedinskaitéa and Akstinaitéa 2014; Mwanza and Chingarande 2013) as well as airport specific models (e.g. Fodness and Murray 2007; Pantouvakis and Renzi 2016). The passenger experience is a newer concept to emerge. It concentrates on areas perceived as significant to passengers, rather than on how services are being delivered, which is the key role of conventional service quality assessment. It involves taking a subjective holistic perspective of the various encounters that passengers face in their airport journey, whereas service quality is more about measuring variables using specific criteria. Therefore, the passenger experience needs to cover the overall door-to-door experience, including transport to/from the airport. Also, a wider viewpoint of service provision is required by considering other organisations, such as airlines and government agencies, as well as the airport operator, that contribute to the entire journey. Ultimately the overall experience is determined by the weakest link, and passengers are rarely familiar or concerned with the distribution of responsibility amongst different service providers. A relatively new way to investigate a more holistic view is through an analysis of online passenger reviews, typically as reported by Skytrax (Bogicevic et al. 2013; Wattanacharoensil et al. 2017) or Google (Lee and Yu 2018).

Within the limited literature in this area, the airport experience has been described as all the activities and interactions that passengers encounter at an airport, with these activities being divided between the necessary processes and discretionary activities (Popovic et al. 2009). It is defined as the net impression of all of the experiences a passenger has in an airport as judged by a passenger's individual standards, expectations and perceptions (Boudreau et al. 2016). Airport experiences have been examined from the three key perspectives of airport management, passenger and the public (Harrison et al. 2012) and by investigating the relationship between time sensitivity and the degree of passenger engagement (Harrison et al. 2015). They also have been considered at three different levels in a so-called pyramid of passenger perception levels (ACI-Europe 2014). At the bottom of the pyramid is the required level covering all basis and mandatory elements. The second level is related to what passengers expect of an airport whereas at the top 'valued' level there are features that surprise passengers and create a 'wow factor'. Progressing up through the different levels of experience is somewhat similar to Maslow's well-known hierarchy of human needs model (Boudreau et al. 2016). Employing state-of the-art technology with the essential processes at an airport is commonly viewed as a key way to enhance the passenger experience (Barich et al. 2015).

Within the broader context of airports contributing to the overall visitor or tourist experience, two roles have been identified, namely as an experience facilitator and an experience provider (Wattanacharoensil et al. 2016). To be a facilitator the airport needs to encourage passengers to co-create their travel experience via social media platforms and have effective internet connections. To be a provider, the airport needs to create a sense of place by providing physical settings and cultural artefacts and activities that reflect the destination. By replicating a destination's traits, the tourism experience can be extended right up to the departure gates (Brilha 2008). Incorporating local natural or man-made attractions into the airport name may help to reinforce the tourism message (Halpern and Regmi 2011). Research has shown that passengers mentally link their airport experience with a destination in three ways: by perceiving their airport experience as a representation of the place they were visiting; by viewing the airport within the context of their perception of the characteristics of the destination; and by comparing their airport experience with tourism promotional messages (Wattanacharoensil et al. 2017).

A 'stress-free airport experience' can also be a significant influence on how welcome visitors feel at a destination. Indeed VisitBritain (2018) found that 37 per cent of visitors felt that this was a very important element, with an additional 44 per cent viewing it as quite important. The resulting total importance score of 81 per cent was not far off the top score of 89 per cent for 'accommodating of tourists'. Overall, the airport experience can be considered as a vital part of the so-called visitor journey framework (Lane, 2007). This framework, which has been adopted by many tourist organisations worldwide, follows the visitor through six key interrelated stages starting at the planning stages and finishing at the return journey and beyond, with the airport experience falling within the 'travel to the destination' stage.

The airport experience is influenced by the range of commercial services and facilities on offer, such as shops and food and beverages. For airport operators, these commercial or 'non-aeronautical' revenues on average make up just under half of their total revenues (Airports Council International (ACI) 2017). When airports ceased to be considered as public utilities but rather as self-sustaining business enterprises, it was clearly recognised that their captive passenger market offered a huge potential for developing such revenues. The other 'aeronautical' revenues are generated by fees charged to airlines. Pressure from government economic regulators and cost-conscious airlines has made it increasingly difficult for airports to grow such revenues and so many airports have focused much of their attention on the non-aeronautical areas (Graham 2009) – sometimes giving rise to accusations that airports are just shopping centres or malls with runways. In this way, airport retail spaces can thus provide visitors with the benefits associated with general shopping centres (e.g. secure, weatherproof and traffic free) but also some perceived drawbacks, such as 'safe, predictable chains' (Wallop 2016).

Responding to changing consumer trends and opportunities provided by technology developments can grow non-aeronautical revenues and arguably enhance the airport experience – at least for technology–savvy passengers (Sevcik 2014; Griffiths 2014). Ensuring that passengers have sufficient time to shop at an airport can bring benefits to the destination by enhancing the visitor experience and at the same time providing this lucrative revenue source for airports (Martín-Cejas 2006). An important distinction that needs to be recognised is whether the mix of departing passengers at individual airports is dominated by outward visitors at the start of their travel experience or whether more passengers are just returning home, as this undoubtedly has an influence on what commercial facilities are most suitable to encourage spend and enhance the experience. Of course, not all passengers wish to shop at airports. Indeed, evidence shows that leisure passengers have a stronger preference to do so than business passengers and poorly planned facilities, interfering with the normal flows of airports, can leave passengers stressed and confused. It has been found that greater passenger satisfaction increases commercial spend (ACI 2016; DKMA 2014) but the relationship between the passenger experience and non-aeronautical facilities is complex and may well go in both directions.

Creating a strong local identity and sense of place experience can extend to the non-aeronautical offer and at the same time encourage passenger spend. Indeed, providing local outlets for passengers suffering from global brand fatigue is a growth area (Assies 2017). The character and culture of the city or country the airport serves can be represented by selling local merchandise or gourmet products or by theming the commercial outlets with images from the city. Moreover, the airport terminal can further act as a destination in its own right by providing conference and meeting facilities and event spaces. These facilities can be shared by passengers, local businesses and other customers (Halpern et al. 2011). Some airports may hold events related to aviation, such as air shows, or unrelated events, such as car races or shows (Prather 2013). However, in spite of these opportunities for an airport to be a destination, it does need to be recognised that for most the bulk of non-aeronautical revenues are still generated in the airside (post-security) part of the airport, which only passengers but not visitors can experience. More rigid security regulations in recent years have encouraged this situation with 85–90 per cent of retail being airside considered to be best industry practice (Steer Davies Gleave 2017).

Clearly the airport's influence extends beyond its actual boundaries by creating jobs and income in the local community as well as generating additional indirect (i.e. associated with the suppliers to the airport) and induced (i.e. associated with the spending of direct and indirect employees) effects. An airport may also encourage catalytic impacts (such as inward investment and tourism) in the local area but frequently these are likely to be more geographically spread, although much depends on the attractiveness of the actual surrounding area. Many airports have expanded beyond the boundaries of the traditional airport business by using neighbouring land for hotels, office complexes, trade centres, light industries, freight warehousing, distribution and logistics centres and business parks (Morrison 2009). As a result of this commercial expansion and diversification, multimodal and multifunctional businesses called airport cities can emerge (Reiss 2007, Perry 2013). If the Airport City continues to develop outwards, the boundaries between the airport and its surrounding urban area may become increasingly blurred, and a new urban form known as an Aerotropolis can appear (Kasarda 2013). Such developments can occur in a planned manner with development initiatives and government/regulatory support, or in a more organic manner, with companies reaping the advantages of agglomeration derived from the productivity benefits from being close to one another and from being located in large labour markets.

Characteristics of London Airports

There are five main airports that serve the London area, namely Heathrow, Gatwick, Stansted, Luton and London City, which are mostly operated by private companies. There are other airports that call themselves a 'London' airport, for example Southend which recently grew its business by offering low cost carrier (LCC) services, and Biggin Hill and Oxford that handle predominantly business jets and general aviation traffic. However, the traffic levels at these additional airports are very small in comparison to the main five.

Table 4.1 presents some key passenger characteristics of the major airports. The two runway Heathrow airport handles nearly double the passenger numbers of the next largest airport (Gatwick) and is unique in having more foreign than UK passengers, and a substantial share of transfer traffic. Heathrow is the principal gateway for many foreign leisure visitors, with the majority of long-haul passengers having no option but to arrive at this airport. Gatwick, Stansted and Luton serve predominately UK leisure passengers. LCCs dominate Stansted and Luton, attracting passengers of lower income. The smallest, somewhat niche, airport is London City airport, situated close to the financial centre of London, which attracts higher income passengers, many travelling for relatively short -distance and -duration business trips. It also has the only business-only flights to New York offered by British Airways and provides many business-focused features such as a Bloomberg hub and hotel/office bag delivery service.

The three largest airports (Heathrow, Gatwick and Stansted) were all owned by the same operator BAA (which was privatised in 1987 with a share flotation) until 2009 when the UK competition authority, formerly known as the Competition Commission, ruled that this common ownership be split up to encourage more competition (Competition Commission 2009). BAA completed its sale of Gatwick in 2009 and BAA's successor entity, Heathrow Airport Holdings (HAH), divested itself of Stansted in 2013. Heathrow and Gatwick are now owned by consortia of investment and infrastructure fund organisations, while Stansted is owned by the Manchester Airport Group (which also owns Manchester and East Midlands airport) with ownership being 64.5 per cent with local Manchester councils and 35.5 per cent with a private investment

	Heathrow	**Gatwick**	**Stansted**	**Luton**	**London City**
Passengers (000s)	75,169	42,146	24,060	14,583	4,501
Type (%): International Domestic Transfer	61 3 36	85 7 8	87 7 6	91 7 2	74 24 2
Purpose (%): Business Leisure	26 74	14 86	14 86	13 87	53 47
Residency (%): UK Foreign	40 60	72 28	64 36	72 28	59 41
Mean trip length (days)	9.6	7.3	5.8	6.8	4.0
Mean income	£55,639	£52,234	£41,682	£39,094	£66,683
Share of all UK inbound visitors (air/sea): Europe (%)	14	13	15	n/a	n/a
Share of all UK inbound visitors (air/sea): Rest of world (%)	62	13	3	n/a	n/a

Table 4.1: Passenger Characteristics at the Major London Airports 2016; n/a = not available but the percentages are smaller than at Heathrow, Gatwick and Stansted. Sources: CAA and International Passenger Survey.

fund. Since its opening in 1987, London City has always been run by private operators. Luton airport was handed over to a private operator in 1998 on a long-term 30-year concession, although the local council maintains ownership. This predominantly private management of all the major airports is relatively rare compared with many other countries, and has meant that achieving a good level of financial performance to satisfy investors has inevitably always been a key priority of the airports.

Largely as a consequence of BAA privatisation in 1987, Heathrow and Gatwick airports are economically regulated and licenced by the UK Civil Aviation Authority (CAA), because it has been determined that they have significant market power which could potentially be abused. At Heathrow there is a price control on the aeronautical fees that are charged to the airlines (CAA 2014a). Meanwhile since 2014 Gatwick airport has been licensed under a so-called Commitments Framework where the CAA monitors a series of commitments on price and other conditions that the airport operator has agreed with its major airlines (CAA 2014b). The other airports are not economically licenced by the CAA and so have total pricing freedom, with Stansted having been deregulated in 2014 after being judged as no longer possessing significant market power (CAA 2013). The regulatory regime at Heathrow airport also covers service quality, as the price control incentives that aim to reduce costs could inadvertently incentivise reductions in service quality. A similar mechanism has been embedded in the Commitments Framework at Gatwick airport. As regards non-aeronautical revenues, all regulated and non-regulated airports are effectively free from any controls, and so are provided with a strong incentive to increase these revenues.

Responding to Service Quality and Passenger Satisfaction Demands

In spite of this regulatory control of service quality, a decade or so ago the overall level of service quality at BAA airports, particularly Heathrow, was perceived as being very poor. Indeed McNeill (2010, 2859) stated that: 'In 2007 London Heathrow airport seemed, by all accounts, to be falling apart' and highlighted the various negative terms such as 'Heathrow Hassle', 'Deathrow', 'Thiefrow', and 'Flightmare' that had been used in the popular press to describe the airport. Moreover Stephens (2007) of the *Financial Times* stated:

> The depressing thing is the relentless predictability of it all. The interminable delays at security. The shuffling crowds in search of somewhere to sit as yet another flight is delayed. Worst, when you eventually escape you know the reprieve is only temporary. You will soon be flying back to broken travelators, long queues at immigration and mayhem in baggage reclaim. Welcome to Heathrow.

Other interested parties, such as the business lobby group London First (2008), went further by declaring that Heathrow's service deficiencies were having a detrimental impact on the attractiveness of London to business and investors. The situation was made worse by the disastrous opening day for the new Terminal 5 in March 2008 when flights were cancelled, luggage delayed and long queues developed. In 2007 within a sample of 101 airports worldwide, in terms of overall passenger satisfaction as measured by the ACI airport service quality (ASQ) survey, Heathrow was ranked a dismal 90th, Gatwick 75th and Stansted 74th. Security provision was a particular area of concern, especially after the introduction of liquid controls in 2006, and the relative rankings regarding securing queues satisfaction were very poor (97th, 93th and 98th respectively) (Competition and Markets Authority (CMA), 2016).

Various possible reasons for such low satisfaction levels (such as airport privatisation with inappropriate profit-maximising objectives, group ownership deterring competition, ineffective economic regulation) were widely debated. Subsequently this poor service quality performance was a major factor, amongst others, in driving some key changes at the airports. As already mentioned, Gatwick and Stansted were both sold to new owners, a new regulatory system was introduced in 2014, and both Gatwick and Stansted entered into long-term agreements with their main airline customers for the first time. At Heathrow the new Terminal 2 was opened in 2014, arguably setting new standards for the passenger experience, and also significant refurbishments to Terminals 3 and 4 have been made. The continuous construction works associated with Terminal 2, and before that with Terminal 5 (opened in 2008), undoubtedly did not help the image of Heathrow. At Gatwick capital investment levels increased after 2009 under the new ownership.

Whilst it is difficult to isolate the combined effects of these developments, and other factors influencing service quality, such as new technology and other operational initiatives, there is considerable evidence to suggest that the overall situation has improved, particularly since the splitting up of BAA (e.g. CMA 2016; ICF 2016; and OXERA 2016). Notably the ACI passenger satisfaction scores for Heathrow and Gatwick have risen considerably (Figure 4.1). In the problem area of security screening, 'very satisfied' passenger ratings at Heathrow increased from 52 per cent in 2008 to 70 per cent in 2015 (Department of Transport (DfT) 2016). By contrast at Stansted, deterioration in overall passenger satisfaction values was observed after 2013. This may well be due to the transfer of ownership in that year, which subsequently led to major investment and re-configuration of its terminal causing temporary significant passenger disruption during this time. CMA (2016) noted that these should, in the longer term, yield service improvement although a snapshot of five key passenger satisfaction measures for July 2017 and 2018 actually shows no improvement (Stansted Airport 2017; 2018).

A comparative assessment of the current service quality and passenger satisfaction levels at the five major London airports is provided with Table 4.2.

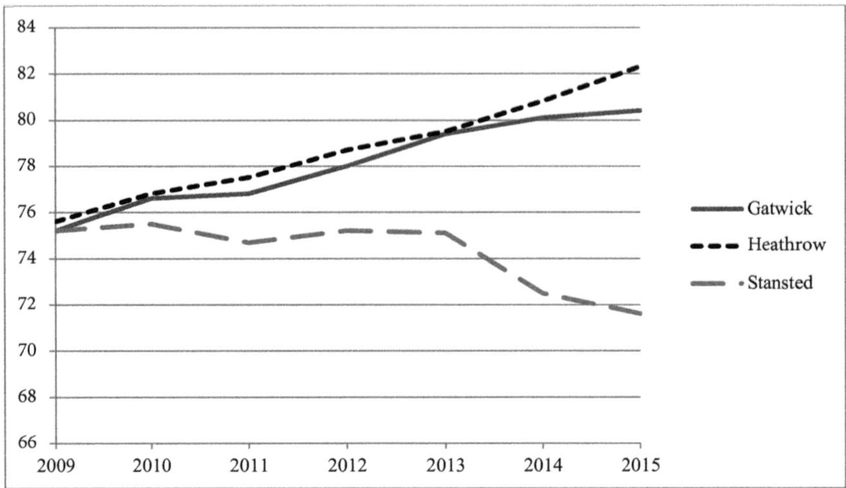

Figure 4.1: ACI ASQ Scores (converted to percentages) at Heathrow, Gatwick and Stansted 2009–2015. Source: CMA (2016).

A survey by the UK consumer association Which? shows highest levels of passenger satisfaction with London City, followed by Heathrow and Gatwick. This confirms the poorer perceptions of Stansted airport with Luton having even lower scores. Fairly comparable views have been found from the CAA airport surveys, although in this case Gatwick fares better. Similar rankings are seen with the Skytrax passenger ratings, which, as mentioned, is a global online air transport review site. Another measure of service quality is the overall delay to the flight. Generally, this correlates with the satisfaction levels, with the notable exception of Gatwick, suggesting that, in assessing overall service quality, passengers may separate experiences within the terminal from temporal ones associated with delays.

In aiming to provide a service primarily for full service carriers (FSCs), London City and Heathrow airports have many more facilities which can, arguably, encourage higher levels of passenger satisfaction. These FSCs will also tend to offer more at the airport because of their business model (e.g. airline lounges, more check-in desks). Passengers flying on LCC services have grown to expect a lower level of service on-board, although this has become more complicated recently by an apparent gradual convergence between the LCC and FSC models. The relevant issue here is whether LCC passenger expectations are lower in relation to the airport service levels as well. The lower satisfaction levels at Stansted and Luton seemingly suggest that this is not the case and some research has confirmed that passengers expect the same airport service whatever fare they pay (ORC International, 2009), even though airports serving LCCs will be under pressure to provide more cost-efficient and simpler facilities to suit

	Heathrow	Gatwick	Stansted	Luton	London City
Which? passenger satisfaction ratings	T5:61% T2:57% T3:52% T4:52%	North 51% South 52%	38%	29%	68%
CAA airport survey passenger experience ratings (% rating it as excellent or good)	89%	91%	82%	83%	91%
Skytrax passenger satisfaction ratings	4/10	4/10	2/10	2/10	5/10
Average flight delay (mins)	13.5	22.3	14.4	18.3	12.8

Table 4.2: Passenger Satisfaction Ratings and Flight Delay at the Major London Airports 2016/17. Sources: Which?, Skytrax and CAA.

the needs of this specific airline model. Moreover, Stansted was not initially designed with this type of traffic in mind and accommodating this traffic (and ensuring that there is appropriate retail and catering) has been a specific challenge and may perhaps contribute towards explaining its lower ranking.

Airports as Shopping and Leisure Destinations

There are a number of varied examples of the London airports acting as destinations in their own right. For example, both Heathrow and Gatwick host family fun days, as well as more specific events such as the recent LEGO tournament at Gatwick. Heathrow has its own permanent cultural space called T5 Gallery, as does Luton with its Gateway Gallery. Stansted has hosted a charity fun run along the runway whereas the Lord Mayor's Balloon Regatta was launched at London City airport in 2017.

Clearly a key role that the airports play is as shopping spaces, particularly Heathrow, being the largest of the London airports. Figure 4.2 shows that Heathrow airport generates a considerable amount of non-aeronautical revenue per passenger compared with other airports, ranking second in Europe (with Gatwick ranked fifth). Overall amongst 50 global airports Heathrow is ranked seventh (LeighFisher 2017). Although such data is very much dependent on traffic mix and passenger characteristics, one way of interpreting this is that the high spend is an indication of good service quality and an attractive range of non-aeronautical services. Alternatively, this larger than average spend could also reflect higher than average non-aeronautical prices. In tracking real net retail revenues per passenger through time (total non-aeronautical revenues in this format being not available) (Figure 4.3), there has been only very limited growth. This reflects the situation at a number of other airports where various external factors such as the economic recession, increased competition from internet shopping and more stringent security measures reducing dwell time, have diminished the ability of airport operators to increase their non-aeronautical revenues. Nevertheless, if the relationship between service quality and non-aeronautical revenue is strong, and given the significant perceived improvements in passenger satisfaction at Heathrow, it is somewhat surprising that there has not been more of an upwards trend in these non-aeronautical revenues.

When passenger satisfaction levels were poor at Heathrow in 2008 it was reported from stakeholder interviews undertaken by London First (2008, 36) that:

> Almost without exception, respondents (and especially those representing business stakeholders) indicated that there was too much space devoted to retail activities to the detriment of core activities directly related to the operation of an airport.

However, given the apparent improvements in survice quality, this may not be a reflection of current views. Some qualitative insight can be gained by looking at the specific Skytrax passenger reviews that focus on the non-aeronautical area

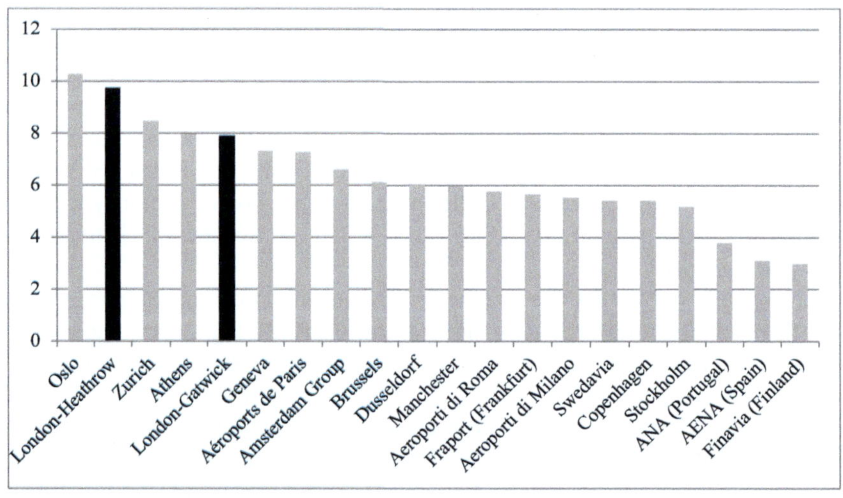

Figure 4.2: Non-Aeronautical Revenue per Passenger 2016 (£s*) Source: LeighFisher (2017).
*converted using Special Drawing Rights (SDR).

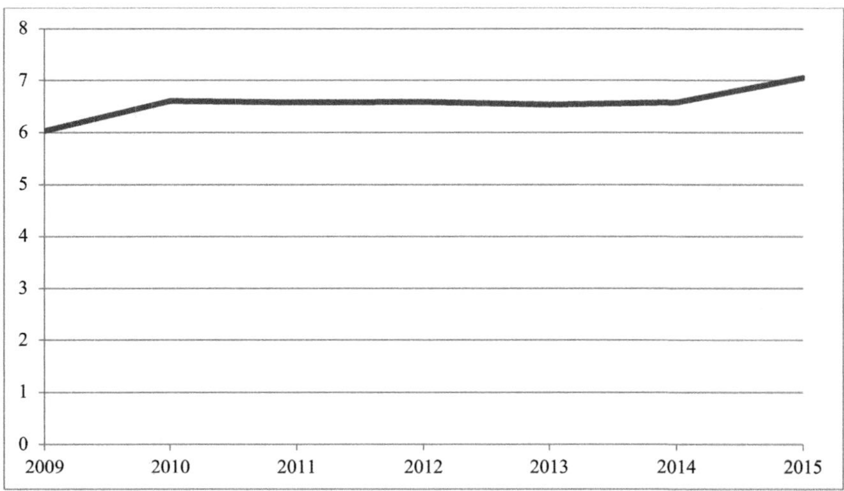

Figure 4.3: Net Retail Revenues per Passenger at Heathrow 2009–2015 (£ – 2015 prices). Source: Steer Davies Gleave (2007).

(although acknowledging that these are voluntary responses from an unrepresentative sample that will tend to pick up only extreme and predominantly negative views) (Appendix 4.A). The varied content of the comments is as expected given passengers' different preferences for shopping, and there are a number of favourable remarks related to the range of facilities and the shopping experience. However, these are outnumbered by negative comments concerning the high prices of goods, the excessive space allocated to retail and its interference with the operational and functional role of the airport. There is some evidence that certain passengers are far from satisfied with the non-aeronautical aspects of the Heathrow airport journey. This mirrors some of the criticism of retail in general in London, where there are thought to be too many shops, significant challenges due to changing technology and shopper habits, and a need for retailers 'to re-think and integrate the experience they offer in physical stores and invest in better customer journeys' (Stevens 2017).

Representing Gateways to London

Having highlighted some important findings related to airport service quality and its role as a shopping destination, this can now be developed further by focusing on the overall passenger experience at Heathrow and its specific role as a gateway to London for inbound visitors. After the dismal service quality ratings of 2007 and 2008, there was increased interest in the overall passenger experience concept, which had previously received little attention. Research at this time emphasised the improvements required in the door-to-door journey experience (DfT 2007), the need for the consideration of a range of softer factors that could not be measured with the typical service-quality measures (Sykes and Desai 2009) and the importance of assessing the interfaces of the different service providers (CAA 2009). A number of these issues were subsequently considered by the airport operator, the CAA, and other interested organisations.

Specifically, in 2009, Heathrow launched a new brand identity with the tagline 'making every journey better' and has continued to promote this message. Initiatives have included introducing multi-lingual 'journey ambassadors' (dressed in the Heathrow brand purple colour), new Central London advertising campaigns, the launch of a mobile app, TV Christmas advertising and other areas of innovation, such as family lanes in security, and more themed events – a recent example being children's activities based on the popular Mr Men books. However, it is difficult to assess the impact of these softer elements of Heathrow's service provision. Two recent comments on Skytrax expressing very contradictory views about the journey ambassadors provide a good example:

> 'What I like most about Heathrow were the staff especially the purple coated helpers whose job is to assist and guide passengers'
>
> 'As for the "purple people" Heathrow staff who seem to do nothing but scowl - what do they actually do?'

Moreover, the journey ambassadors were subject to a recent UK TV documentary that heavily criticised the fact that they were paid commission to direct passengers to the airport's shops with sales targets to meet (around £2,500 a day: Ellson 2017). Such reporting clearly reinforces the views expressed by some in the Skytrax reviews that the airport focuses too much on exploiting its retail opportunities.

In terms of specifically helping inbound visitors, the airport operator has developed a part of its website called 'First Time in Britain' where it has information about the language, weather, driving, doing business, opening hours, public holidays, money, time zones, distances/measurement/sizes, food and drink, tipping, electricity and telephones. Details of surface transport links are provided, although again there has been some criticism of the airport's push to grow revenues, by very much focusing on its own Heathrow Express train service in contrast to the cheaper Heathrow Connect option which is jointly provided by the airport operator and the Great Western Railway train operator (CAA 2016). Like many airports, Heathrow also organises joint promotional activities and other events with the travel trade. A recent example in 2017 was a joint initiative with VisitBritain as part of their GREAT Britain campaign to build awareness of Britain's attractiveness as a tourism destination by showing over 200 artworks. These displayed a mix of heritage sites, pioneering British businesses, and cultural attractions across the UK.

Assessing the impact of such initiatives is hard to measure. The Skytrax comments, whilst again acknowledging the methodological shortcomings, can nevertheless provide some qualitative insight (Appendix 4.B). The comments indicate that these passengers feel the airport is an integral part of the visit or holiday experience and should be providing an ambassadorial role for London. However, the reality seems to be that a number of these passengers were left with a less than favourable impression and welcome.

Anchors for Wider Development

Heathrow, as with most airports, has a significant role in the wider local community. According to the latest employment survey, it generates more than 75,000 direct jobs at the airport with around 54 per cent of these coming from the five closest local London boroughs that border the airport site (Hounslow, Hillingdon, Ealing, Slough and Spelthorne) (Heathrow Airport Ltd 2014). The direct local jobs account for around 16 per cent of total local employment in this area and if the indirect and the induced jobs are added in, this accounts for up to 22 per cent of local jobs (Optimal Economics 2011). The catalytic or spin-off impacts related to tourism, or the role of the airport in facilitating tourism, are much more widely felt in London and beyond. The actual local catalytic impacts at Heathrow are very difficult to quantify (PWC 2014) and there is no certainty as to whether all local tourism opportunities are really exploited, as the following Hillingdon Council (2007, 40) statement (albeit more than ten years ago) suggested:

> visitor perception to the area is very limited, with the result that visitors are missing out on opportunities to visit and experience other places and services.

The local area around Heathrow can also be attractive for conferences and meetings (and indeed other events) as explained in the promotional message of a meetings organiser (meetingpackage 2017):

> With more than 27 million business travellers in 2015, the airport offers a great opportunity to … save some of your precious time. And money. Just give it a thought. Why hold a meeting or a conference with more than 50 people from all around the world (or not) in a place out of the airport where they initially land? Your clients or business colleagues will already be there. You'll avoid additional costs, traffic, save some time and be productive.

Whilst demand-side data related to this is scarce, a quick internet search reveals some indication of the level of supply. The conference organiser *Conferences UK* had venue partnerships in place with 18 meeting venues in the Heathrow area with over 51,000 meeting rooms to choose from, while the *Venue Directory* had 33 matches for the vicinity.

Many of these facilities are offered by accommodation providers, which are one of the key sectors in the local community that benefit from Heathrow's presence. The area around Heathrow has around 150 accommodation establishments accounting for over 15,000 rooms and making up 11 per cent of the total accommodation stock in London (London and Partners 2017). Traditionally many of these tended to be rather exclusive and upmarket but with factors such as changing lifestyles, cheaper long-distance travel and the growth of budget hotel chains, there is now much more cheaper accommodation available (as in other parts of London). In Hillingdon, there were 42 hotels (9,701 rooms) in 2016 and although 64 per cent of these were 4–5 star, budget hotels represented a significant proportion (19 per cent) (London and Partners, 2016), with notable new additions to the stock including IBIS Styles (140 rooms) in 2016 and the Premier Inn (613 rooms) in 2017.

The majority of the hotels and their facilities are designed to meet the needs of passengers (both origin/destination and transfer) using the airport, but they can also be used for local demand (e.g. VFR, business, and for special events in West London). They are a necessary condition for the development of both the local tourism and events industries. In addition, Heathrow has traditionally been one of the main overspill accommodation areas when central London is full but the local hotels now face more competition with the growth of the cheaper hotel stock in the outskirts in other areas of London, particularly in the East. Much of London's development is in an easterly direction and with the development of key visitor attractions close to areas such as Queen Elizabeth Olympic Park and Stratford, together with the popularity of venues such as the O2 Arena and ExCeL centre (and closure of Earls Court in the West), spreading the London destination westwards is challenging.

There may be visitor perception issues as identified by the Royal Borough of Windsor and Maidenhead (RBWM) (2017, 15):

> There is a risk that some overseas visitors are deterred from staying in RBWM because they perceive it to be easier to travel into Central London, losing potential revenue. There is a concern that visitors from overseas (including those coming to the UK for business) do not appreciate the proximity of Windsor to Heathrow (7.2 miles) and consider it easier to use London hotels as a base. This suggests an opportunity to reposition the destination to make its proximity to London much clearer.

As a result of its tourism plan of 2017-2020, Windsor and Maidenhead propose a rebrand to ensure that its proximity to London and Heathrow is fully understood. Heathrow's surface transport connectivity will be improved significantly with the opening of the new east–west Elizabeth Line (also known as Crossrail) which will replace the existing Heathrow Connect service probably in 2020, although this is will be over a year later than originally planned with a substantially higher cost. This may enhance Heathrow's position as a gateway to London by making it easier to directly access more parts of London and in turn may stimulate more catalytic development. Arguably this could have a positive or negative impact on tourism and events in the local area, on the one hand making it easier for people to stay in the area and travel into London, but on the other hand making it easier to bypass the area altogether. In the longer term a connection with the planned high-speed HS2 via an interchange with the Elizabeth Line may also be an option.

It is also worth noting that because of Prince Harry's and Princess Eugenie's weddings at Windsor in 2018, Windsor and the surrounding area experienced a considerable boost to its tourism industry which may linger afterwards. Indeed, on the tourism page of the Royal Borough of Windsor and Maidenhead website it is stated: 'If you were inspired by what you saw on television then come and experience the Windsor Welcome for yourself! There has never been a better time to visit!' (Royal Borough of Windsor and Maidenhead 2018). Even nearby Slough, close to Heathrow but 'not a typical tourist hotspot' may benefit from the extra exposure that it received at a time when there has been considerable regeneration investment for the town (Fishwick 2017).

There is evidence of new development being planned around Heathrow airport which may enhance the wider economic impacts. For example, the West of Borough area of Hounslow was designated by the Mayor of London in 2015 as a so-called Opportunity Area for significant business growth and housing development with improved public transport access. Within this area, Hounslow Council has proposed the development of a new urban quarter which will include Heathrow Garden City with new homes, a Heathrow Gateway business hub and an Airport Business park which is currently under consultation.

Overall, in acknowledging the impact that Heathrow airport has on the local community, this raises the issue as to whether Heathrow can be considered as an Airport City or Aerotropolis. A clustering of hotel and conference facilities, together with a concentration of business activity in areas such as Stockley Park and Uxbridge Business Park, suggested to Kasarda (2013) that it could be. However, this development has occurred in an organic fashion (as indeed has the development of the airport), which is in stark contrast to Manchester where a different local and regional context has led to the development of the UK's first planned Airport City. This has a total area of 500,000 m^2 to be occupied eventually with offices, hotels, advanced manufacturing, logistics facilities, and ancillary retail space (Manchester Airport Group 2017).

Finally, in considering the implications for wider development, mention must be made to plans to expand the capacity of the London airports. Undoubtedly the most important issue here has to be where to build an additional runway which has been fiercely debated over many years. Most recently, the independent Airport Commission considered all the options, narrowed this down to two at Heathrow (an additional runway to the north west of the existing runways or a new extended runway to the west of the existing northern runway) or one additional runway at Gatwick, and eventually recommended the new additional north west runway at Heathrow (Airports Commission 2015). The UK government has accepted this recommendation as part of the Airports National Policy Statement (DfT 2018).

The other major London airports are also planning expansion, with a significant development at London City airport providing for two million more passengers by 2025. Originally these plans had been opposed by the previous mayor Boris Johnson but the current mayor, Sadiq Khan, controversially dropped this objection shortly after being elected in 2016 – even though he had pledged to be London's 'greenest' ever mayor. Luton airport published growth plans in December 2017 which could double the number of passengers by 2050. The proposals are still subject to planning permission but include a so-called New Century Park aiming to encourage business development and employment opportunities, together with enhanced community facilities, within the neighbourhood of the airport. Stansted airport is also planning expansion although needs to get planning permission to raise the cap on annual passengers.

Conclusions

This aim of this chapter was to consider the role of London's airports as experiences and destinations. It is argued that the London airports can act as:

1. Destinations in themselves where people do more than simply use transport facilities (e.g. they use retail and leisure facilities).
2. Gateways to London, providing the first impression of London and the first part of this destination experience.
3. Anchors for the development of a wider destination area (e.g. Windsor and Maidenhead).
4. Transport nodes that can assist the development of nearby surburban destinations (particularly with new links such as the Elizabeth Line and HS2).

Looking forward, there are a number of key factors that are likely to have a major impact on London airports and the subsequent balance of these four dimensions. Undoubtedly the most significant of these will be the development of a third runway at Heathrow. Although in principle this has been approved, a considerable amount of opposition remains, particularly on environmental grounds, which is likely to lead to legal challenges associated with this government decision. If these are unsuccessful, the earliest possible date for the third runway would be 2026.

If there is this increase in runway capacity, Heathrow's ability to perform its role as the primary gateway to London could be enhanced, with airlines more able to develop new routes, especially from long-haul destinations in emerging economies with growing tourism demand which have been squeezed out of Heathrow at the moment. If, for whatever reason, the runway is not built, which is not totally improbable given past failed attempts, Heathrow's ability to position itself as one of the world's global hubs is likely to be challenged. Actually, this is already happening, not only because of capacity constraints, but also as the major air transport markets move eastwards and consequently make Middle Eastern and Asian hubs more popular. While a weakening of a hub function is likely to affect the feasibility of certain routes, it is difficult to conceive that Heathrow will not remain the major origin and destination airport for visitors to London, even though lack of capacity could cause some displacement of leisure passengers to other airports, with business visitors being prepared to pay a premium for access to Heathrow. This could then have an impact on other smaller London airports, perhaps putting more pressure for the need for service quality enhancements, particularly at Stansted and Luton, where currently passenger satisfaction appears to be relatively low.

Meanwhile, by contrast, most evidence points to improvements in service quality levels at both Heathrow and Gatwick over the last ten years. Moreover, an interesting proposed change for 2020 onwards, announced by the CAA, will be the way that service quality is regulated at Heathrow. The plan is to shift to an outcome-based approach that focuses on considering what airports are actually delivering to users rather than how they deliver it – hence moving closer to the passenger experience concept (CAA 2017).

The forces explaining the differences in service quality at the different London airports are undoubtedly complex. Privatisation, with a profit maximising objective, is an obvious factor, but all five major airports have been under private management for some time. Some (Heathrow, and Gatwick/Stansted up until recently) have had their service quality officially regulated whereas the unregulated airports (London City and Luton) are ranked top and bottom and so it is difficult to detect a causal relationship here. Evidence does, however, suggest that the splitting up of BAA has played a major role in improving quality standards at Heathrow and Gatwick as the airports face more competition. On the other hand, the lower–performing Stansted and Luton airports likewise operate in a fairly competitive environment, which suggests their focus on serving the LCC airline model may play a key role here in influencing service levels.

The Skytrax reviews indicate that many passengers are unhappy with the non-aeronautical facilities or shopping experience at Heathrow. One possible way to increase passenger satisfaction could be to fully embrace the use of technology and simplify the retail experience, especially with navigational tools for shops, and the availability of online purchases and ordering. Such developments are in line with general retail trends and have been partially embraced by Heathrow airport but potentially could be exploited more. This could perhaps reduce some of the criticisms which relate to shopping physically dominating the airport experience. The reviews also suggest that Heathrow is not always viewed as a good ambassador for London. It is noteworthy that many of the airport's initiatives such as loyalty programmes, car parking packages with hotels and events like the Mr Men initiative are much more attractive to outbound UK residents, rather than inbound visitors, which is possibly where more consideration could be given, especially as UK residents only account for 40 per cent of the traffic. As discussed earlier, the relative mix of inward and outward passenger flows must be a key consideration when commercial services and activities are being planned.

In looking to the future, it is clear that the Heathrow third runway, but also other expansion plans at the other airports, offer significant opportunities to enhance the visitor experience in the long term, although the more short-term disruption needs to be very carefully managed. Inevitably all these plans are being vehemently opposed by certain resident groups and environmentalists but, if they all do go ahead, they may have the ability to unlock additional capacity and help a bit to disperse the benefits that airports can bring as destinations, gateways and anchors for development around a greater area of London.

References

ACI. 2016. *Does Passenger Satisfaction Increase Airport Non-aeronautical Revenues?* Montreal: ACI.

ACI. 2017. *Airport Economics 2017 Report.* Montreal: ACI.

ACI Europe. 2014. *Guidelines for Passenger Services at European Airports.* Brussels: ACI Europe.

Airports Commission. 2015. *Final Report.* London: Airports Commission

Appold, Stephen and John Kasarda. 2011. 'Are Airports Non-Places?' *Airport Consulting*, Summer, 15–17.

Assies, Hildegard. 2014. 'Local Heroes on the Move: How a Growing Number of Local Brands are Gaining Presence at Airports'. *Journal of Airport Management*, 8(4), 305–307.

Augé, Marc. 1995. *Non-Places. Introduction to an Anthropology of Supermodernity* (English translation of *Non-Lieux. Introduction à une anthropologie de la surmodernité* by John Howe). London: Verso.

Barich, Frank, Ruiz, Leslie, and Jim Miller. 2015. 'Enhancing the Passenger Experience through an Integrated Approach to Self-Service Opportunities'. *Journal of Airport Management*, 10(1), 49–63.

Bogicevic, Vanja et al. 2013. 'Airport Service Quality Drivers of Passenger Satisfaction'. *Tourism Review*, 68(4), 3–18.

Boudreau, Bruce et al. 2016. *ACRP Report 157: Improving the Airport Customer Experience*. Washington, DC: Transportation Research Board.

Brilha, Nuno Mocica. 2008. 'Airport Requirements for Leisure Travellers'. In Anne Graham, Andreas Papatheodorou and Peter Forsyth, eds. *Aviation and Tourism: Implications for Leisure Travel*. Aldershot: Ashgate.

CAA. 2009. *The Through Airport Passenger Experience: An Analysis of End-to-end Journeys with a Focus on Heathrow*. London: CAA.

CAA. 2013. *Market Power Determination for Passenger Airlines in Relation to Stansted Airport – Statement of Reasons*, CAP 1135. London: CAA.

CAA. 2014a. *Economic Regulation at Heathrow from April 2014: Notice granting the Licence*, CAP 1151. London: CAA.

CAA. 2014b. *Economic Regulation at Gatwick from April 2014: Notice granting the Licence*, CAP 1152. London: CAA.

CAA. 2016. *Review of Market Conditions for Surface Access at UK Airports – Final Report*, CAP 1473. London: CAA.

CAA. 2017. *Guidance for Heathrow Airport Limited in Preparing its Business Plans for the H7 Price Control*, CAP 1540. London: CAA.

CMA. 2016. *BAA Airports: Evaluation of the Competition Commission's 2009 Market Investigation Remedies*. London: CMA.

Competition Commission. 2009. *BAA Airports Market Investigation*, Final Report. London: Competition Commission.

DfT. 2007. *Improving the Air Passenger Experience*. London: DfT.

DfT. 2016. *Air Passenger Experience of Security Screening: 2015*, Statistical Release, 8 September. London: DfT.

DfT. 2018. *Airports National Policy Statement: New Runway Capacity and Infrastructure at Airports in the South East of England*. London: DfT.

DKMA. 2014. *Passenger Satisfaction: The Key to Growing Non-Aeronautical Revenue*. Available at http://www.dkma.com/en/images/downloads/commercial/Passenger%20satisfaction%20-%20the%20key%20to%20growing%20non-aeronautical%20revenue.pdf (accessed 4 July 2017).

Ellson, Andrew. 2017. 'Heathrow Helpers "Just Want Your Money"' *The Times*. Available at https://www.thetimes.co.uk/article/heathrow-helpers-just-want-your-money-q8v2dbmcm (accessed 1 December 2017).

Fishwick, Samuel. 2017. 'Slough is on the Up (No, Really): How the Royal Wedding will Boost this Suburb with Superpower'. *The Evening Standard*. Available at https://www.standard.co.uk/lifestyle/london-life/slough-meghan-markle-prince-harry-royal-wedding-a3707526.html (accessed 10 February 2018).

Fodness, Dale and Brian Murray. 2007. 'Passengers' Expectations of Airport Service Quality'. *Journal of Services Marketing*, 21(7), 492–506.

Gottdiener, Mark. 2001. *Life in the Air: Surviving the New Culture of Air Travel*. Oxford: Rowman and Littlefield.

Graham, Anne. 2009. 'How Important are Commercial Revenues to Today's Airports?'. *Journal of Air Transport Management*, 15(3), 106–111.

Griffiths, Stéphanie. 2014. 'The Future of Airports: Capitalising on Mobile Devices to Enhance the Traveller Experience and to Maximise Retail Opportunities', *Journal of Airport Management*, 8(4), 312–317.

Halpern, Nigel and Uttam Regmi. 2011. 'What's in a Name? Analysis of Airport Brand Names and Slogans'. *Journal of Airport Management*, 6(1), 63–79.

Halpern, Nigel, Graham, Anne, and Rob Davidson. 2011. 'Meeting Facilities at Airports'. *Journal of Air Transport Management*, 18(1), 54–58.

Harrison, Anna et al. 2012. 'Challenges in Passenger Terminal Design: A Conceptual Model of Passenger Experience'. *Proceedings of the Design Research Society (DRS) 2012 Conference*, Chulalongkorn.

Harrison, Anne, Popovic, Vesna, and Ben Kraal. 2015. 'A New Model for Airport Passenger Segmentation'. *Journal of Vacation Marketing*, 21(3), 237–250.

Heathrow Airport Ltd. 2014. *Taking Britain Further: Connecting all of the UK*, Technical Submission to the Airport Commission, Vol 1. London: Heathrow Airport Ltd.

Hillingdon Council. 2007. *Local Development Framework Background Technical Report Tourism Study*. Hillingdon: Hillingdon Council.

Honegger, Matt. 2013. 'Culture and Function: "Airports have transformed from transportation hubs into shopping malls"', *Airport World*. Available at http://www.airport-world.com/item/2182-culture-and-function (accessed 15 February 2018).

Huang, Wie-Jue, Honggen Xiao and Sha Wang. 2018. 'Airports as Liminal Space'. *Annals of Tourism Research*, 70, May: 1–13.

ICF. 2016. *Evaluation of the 2009 Competition Commission's BAA Airports Market Investigation Remedies*. London: ICF.

Kasarda, John. 2013. 'Airport Cities: The Evolution'. *Airport World*, April–May, 24–27.

Lane, Mandy. 2007. 'The Visitor Journey: The New Road to Success'. *International Journal of Contemporary Hospitality Management*, 19(3), 248–254.

LeighFisher. 2017. *Airport Performance Indicators 2017*. London: Leigh-Fisher.

Lee, Kiljae and Chunyan Yu. 2018. 'Assessment of Airport Service Quality: A Complementary Approach to Measure Perceived Service Quality Based on Google Reviews'. *Journal of Air Transport Management*, 71, 28–44.

Lloyd, Justine. 2003. 'Airport Technology, Travel and Consumption'. *Space and Culture*, 6(2): 93–109.

London and Partners. 2016. *London Hotel Development Monitor*. London: London and Partners.

London and Partners. 2017. *Tourism Report 2015–2016*. London: London and Partners.

London First. 2008. *Imagine a World Class Heathrow*. London: London First.

Losekoot, Erwin. 2015. 'Airports: Places or Non-Places – Who Cares'. *2nd Symposium of the Transport and Tourism Special Interest Group*, Auckland University of Technology, Auckland, 15–17 April.

Lloyd, Justine. 2003. 'Airport Technology, Travel, and Consumption'. *Space and Culture*, 6(2), 93–109.

Manchester Airport Group. 2017. *Airportcity*. Available at http://www.airportcity.co.uk/uploads/airportcity-internationalproductcatalogue.pdf (accessed 15 October 2017).

Martín-Cejas, Roberto Rendeiro. 2006. 'Tourism Service Quality Begins at the Airport'. *Tourism Management*, 27(5), 874–877.

McNeill, Donald. 2010. 'Behind the "Heathrow Hassle": A Political and Cultural Economy of the Privatized Airport'. *Environment and Planning A*, 42(12), 2859–2873.

Meetingpackage. 2017. *12 Excellent meeting room venues near Heathrow Airport.* Available at https://meetingpackage.com/blog/12-excellent-meeting-rooms-venues-near-heathrow-airport-london (accessed 15 October 2017).

Morrison, William. 2009. 'Real Estate, Factory Outlets and Bricks: A Note on Non-Aeronautical Activities at Commercial Airports'. *Journal of Air Transport Management*, 15(3), 112–115.

Mwanza, Cosmas and George Chingarande. 2013. 'An Evaluation of Passenger Satisfaction on Service Quality Delivered at an Airport in Namibia: A Quality Perspective Approach'. *International Journal of Management, IT and Engineering*, 3(1), 292–304.

Oldenburg, Ray. 1989. *The Great Good Place: Cafés, Coffee Shops, Community Centers, Beauty Parlors, General Stores, Bars, Hangouts, and how they get you through the Day.* New York: Marlowe.

Optimal Economics. 2011. *Heathrow Related Employment.* London: Optimal Economics.

ORC International. 2009. *Research on the Air-Passenger Experience at Heathrow, Gatwick, Stansted and Manchester Airports.* London: ORC International.

OXERA. 2016. *Evaluation of the Competition Commission's BAA Airports Market Investigation.* London: Oxera.

Pabedinskaitė, Arnoldina, and Viktorija Akstinaitė. 2014. 'Evaluation of the Airport Service Quality'. *Procedia-Social and Behavioral Sciences,* 110, 398–409.

Pantouvakis, Angelos and Maria Francesca Renzi. 2016. 'Exploring Different Nationality Perceptions of Airport Service Quality'. *Journal of Air Transport Management,* 52, 90–98.

Popovic, Vesna, Kraal, Ben, and Philip Kirk. 2009. 'Passenger Experience in an Airport: An Activity-Centred Approach'. *IASDR 2009 Proceedings,* Seoul.

Prather, Daniel. 2013. *ACRP Synthesis 41: Conducting Aeronautical Special Events at Airports.* Washington DC: Transportation Research Board.

PWC. 2014. *Local Economy: Literature Review,* a Report for the Airports Commission. London: PWC.

Reiss, Brett. 2007. 'Maximising Non-Aviation Revenue for Airports: Developing Airport Cities to Optimise Real Estate and Capitalise on Land Development Opportunities'. *Journal of Airport Management*, 1(3), 284–93.

Royal Borough of Windsor and Maidenhead (RBWM). 2017. *Tourism Action Plan 2017–2020*. Windsor: RBWM.

Royal Borough of Windsor and Maidenhead (RBWM). 2018. *Windsor and the Royal Wedding*. Available at https://www.windsor.gov.uk/royalwedding (accessed 1 August 2018).

Stansted Airport. 2018. *Service Quality Commitments* July 2017 and July 2017. Available at https://www.stanstedairport.com/about-us/london-stansted-airport-and-mag/our-performance/customer-service/ (accessed 10 October 2018).

Sevcik, Thomas. 2014. 'The End of Retail, the Future of Retail'. *Journal of Airport Management*, 8(4), 308–311.

Steer Davies Gleave (SDG). 2017. *Heathrow Airport – Review of Commercial Revenues*. London: SDG.

Stephens, Philip. 2007. 'Put an End to the Heathrow Misery', *Financial Times*, 11 June. Available at http://www.ft.com (accessed 15 October 2017).

Stevens, James. 2017. 'Leisure will be Key to the Future of London Retail', *City A.M*, 14 October. Available at http://www.cityam.com/265587/leisure-key-future-london-retail (accessed 16 February 2018).

Sykes, Wendy and Philly Desai. 2009. *Understanding Airport Passenger Experience*. London: Independent Social Research.

VisitBritain. 2018. 'Understanding Welcome', *Foresight* – Issue 159. London: VisitBritain.

Wallop, Harry. 2016. Brent Cross is now 40 Years Old. Will Shopping Centres be here in another 40?', *The Telegraph*. Available at http://www.telegraph.co.uk/business/2016/03/01/brent-cross-is-now-40-years-old-will-shopping-centres-be-here-in/ (accessed 16 February 2018).

Wattanacharoensil, Walanchalee, Markus Schuckert and Anne Graham. 2016. 'An Airport Experience Framework from a Tourism Perspective'. *Transport Reviews*, 36(3), 318–340.

Wattanacharoensil, Walanchalee, Markus Schuckert and Anne Graham. 2017. 'An Analysis of the Airport Experience from an Air Traveler Perspective'. *Journal of Hospitality and Tourism Management*, 32, 124–135.

Appendix 4.A: Skytrax Passenger Comments About Non-aeronautical Revenues.

POSITIVE	NEGATIVE
RANGE OF SHOPS	
'On the way out I was astonished by the range of shops after passing security.' 'Lots of shopping and a few coffee shops with some restaurants mixed in.' 'There were plenty of shops inside such as restaurants and duty free.' 'There were more than enough options for eating and shopping at the airport.' 'There are plenty of shops airside if you wanted to make some extra purchases before you flight. There are also plenty of eateries.' 'Well-equipped for shopping and restaurants.' 'Plenty of choice for food because there were so many restaurants and shops.' 'The airport has an excellent selection of shops and restaurants.' 'Brilliant selection of food and retail outlets.' 'Food and duty free selection is impressive.' 'Excellent facilities for shopping and eating.'	'The airport is more like a big mall rather than an airport.' 'I am fed up with the ridiculous amount of expensive stores that clutter the entire terminal and cause traffic bottlenecks.' 'Airports are not shopping centres. It's a shame that LHR is only interesting in profiteering through rent from retailers.' 'Why anyone would rate an airport high because the shopping is good baffles me.' 'Who goes to an airport to shop?'.
TYPES OF SHOPS	
	'Shopping is too high end.' 'Still find it strange this airport wins an award for shopping. For the normal commuter or traveller it only has high brand companies in T2 and T5 which are the most up to date terminals.' 'It's all very well having luxury handbags, jewellery and other high value items available, but does it really appeal to the majority of passengers?'. 'Shopping is pretty dismal too – I don't understand all those empty designer stores'. 'Why not cater for ordinary people with ordinary budgets?'.

POSITIVE	NEGATIVE
	'The shopping seemed far too expensive for me even to venture into some of the stores'. 'Travellers from certain countries may have money to indulge in luxury items, but I believe most would want a last minute memento of their holiday knowing they wouldn't face a huge credit card bill when returning home'.
THE SHOPPING EXPERIENCE	
'I also love the shopping features and terminal layouts'. 'Terminal is spacious and has good selection of shops and eatery's airside'. 'That said our dining experience was utterly superb'. 'Loads of shops/places to eat but still never feels crowded and noisy even at peak times'. 'Was a fine experience with good shopping possibilities'. 'Duty free and departure areas nice and open plenty of food and shops everywhere'. 'Terminals are clean well decorated and for the most part have excellent shops!'.	'Overall my thoughts of Heathrow is that they want my money but forget what the passenger actually wants or how we wish to be treated'. 'This terminal is starting to feel and looks like a Shopping Mall'. 'The airport management at senior level should decide what is more important a stress-free passage for passengers through security or to continue to open more retail outlets. At the moment it appears that the emphasis is to get passengers to spend more at the retail outlets than to establish a stress-free passage through security'. 'Too much like a shopping mall and always unpleasantly overcrowded and noisy'. 'Policy on using most of the space for shopping really does make using LHR terminals a depressingly uncomfortable experience'.
PRICES	
	'However the duty free area is just another gilded souk. Every time that you try and stop to have a look at something, an aggressive sales shark will pounce upon you. I was stopped 11 times'. 'Should you have the misfortune of having to buy something, you are harassed into buying other products. It's an absolute headache and a very unpleasant experience. They lost money from me. Staff are not even qualified in their concessions. Clearly everyone is on targets and minimum wage. It's horrible'.

POSITIVE	NEGATIVE
	'Seeing the amount of space the airport authorities devoted to shopping versus passenger security clearly you consider passengers cash cows'.
COMFORT	
'Large variety of restaurants and cafes and also has a spacious terminal'. 'The shopping area was pleasant and we didn't have trouble finding somewhere to sit'.	'The Duty Free area is way too large and the prices not at all that competitive: area should be halved for more comfort and seating places'. 'There are more shops than are really needed, and for passengers less places to sit and rest before a long flight'. 'They want passengers up and spending money. I found the seats to be not very comfortable and less uncomfortable the longer you sat on it'. 'There is now so much focus on retail areas with little or no thought for the average traveling passenger. …so much space is now taken for retail that there is little space to sit down without feeling overcrowded'.

Appendix 4.B: Skytrax Passenger Comments about the Passenger Experience

POSITIVE	NEGATIVE
ROLE WITHIN THE OVERALL VISIT	
'It's a really lovely way to start a holiday'. 'All in all a very pleasant start to our holiday'.	'Not a good way to start our vacation'. 'The start of our holiday has been a shambles with no apology'. 'A bad experience all around and a lousy end to a holiday'. 'It was an appalling experience that spoilt the start of my holiday of a lifetime'.

POSITIVE	NEGATIVE
WELCOME TO LONDON	
	'It is the primary Gateway to London and Britain. What an awful welcome'.
	'What a way to be welcomed to London and the UK!'.
	'Dirty crowded expensive and inefficient. A microsome of most things in London. Nicely reflects what to expect from England as a visitor'.
	'Did I mention the airport is dirty? I have noticed on the London subway that the locals just leave trash everywhere but in trash cans and they do the same at Heathrow'.
	'The passenger's first taste of London and it leaves an unsavory taste for the entire country'.
	'I was shocked to see this attitude from someone who was my first point of contact with London'.
WELCOME TO THE UK	
'Overall had a very satisfying experience there and a fantastic welcome to the UK'.	'Everyone was tired and then they are subjected to this - a far cry from a good welcome to the UK'.
	'Heathrow is supposed to be the gateway to our country and should show respect and be pleasant to travellers and not treat them like cattle'.
	'I did not feel welcome at all to the UK with that queue and do not recommend coming back!'.
	'Clearly woefully understaffed – what a welcome to the UK for foreign visitors'.
	'My husband, who'd not been to England before, swore never to go there.'
	'It makes me think twice about spending my tourist dollar in a country that apparently just doesn't care'.

POSITIVE	NEGATIVE
IMPRESSIONS	
	'What a bad impression it must give visitors to the UK'. 'The old saying "You never get a second chance to make a first impression" is unknown by Heathrow'. 'What sort of advertisement is that for a visitor to "Great" Britain?'. 'Just imagine what impression this gives visitors to this country the moment they arrive here'. 'Welcome to Britain. No matter how much money the English Tourist Board spends on promotion it's the initial impression that counts'. 'What impression do they give to overseas travellers to our country?'. 'Why can't we get basic courtesy right? First impressions count'. 'Place is a joke. As a Brit, it's frankly embarrassing'. 'I am embarrassed for what visitors will think of Britain'. 'An embarrassment to be British if this is people's first impression of the UK'. 'Few airports are fun but this really gives a poor impression of this country to visitors'. 'I truly feel sorry for tourists entering England for the first time and having to encounter the human zoo'.

CHAPTER 5

The City of Sport: London's Stadiums as Visitor Attractions

Claire Humphreys

Introduction

Sport tourism has grown in academic prominence in recent years, with recognition that stadiums and other sporting locations can contribute to both the tourism offer and to a city's image. This chapter focuses on the major sporting venues in London and examines their appeal as attractions for tourists. Much research has considered the significant role of sport in urban economies (Gratton and Henry 2001) but Higham (2005, 239) goes further, arguing that 'conceptualizing sport as a tourist attraction' can generate revenue from markets seeking a different type of authentic experience. City tourism now encompasses a broader range of facilities and experiences, and sports tourism in London provides an illustrative example of this type of expansion.

Sports tourism includes active participation in sport as well as spectating at sports matches and competitions (Weed and Bull 2004). It also includes viewing sports heritage and places, including museums and halls of fame (Gibson 1998). Consequently, the range of sporting infrastructure which may be used by tourists is extensive. Gammon and Robinson (1997) assessed the relative

How to cite this book chapter:
Humphreys, C. 2019. The City of Sport: London's Stadiums as Visitor Attractions. In: Smith, A. and Graham, A. (eds.) *Destination London: The Expansion of the Visitor Economy.* Pp. 91–116. London: University of Westminster Press. DOI: https://doi. org/10.16997/book35.e. License: CC-BY-NC-ND

importance of sport over other travel motivations and suggested a distinction between those for whom the sporting element dominates and those focused on tourism aspects, with sporting activity included only as secondary aspect. In most cases tourists engage with sports facilities primarily developed to serve a local market but some sports facilities have developed – and promoted – touristic elements to enhance their commercial appeal. Accordingly, sports locations can become sites of tourist consumption (Ginesta 2017) and sporting venues that actively seek tourists may manipulate both the design of the venue and its promotion to maximise tourist revenues.

To ensure the commercial viability of such stadiums, maximising operational revenue is important and tourism provides an additional income stream. Income is generated through event ticket sales, but also through sales of tours of the facilities or through museum visits. Redmond (1973, 45) acknowledges that 'the commercial value of sport as a tourism asset has long been recognised,' and it is worth in excess of £2 billion to the UK economy, evenly split between sports participants and sports spectators (of events and tours) (DCMS 2012).

Iconic spaces such as sporting venues can also play a wider role in destination promotion and other policy objectives: 'Sport is increasingly seen as a central strategy for cities to promote their image and global position, undertake regeneration, and tackle problems of social exclusion' (Herring 2004, 17). For many decades, sports stadiums have been used as catalysts for urban and social development, to capitalise on mass interest in sport (Stevens and Wootton 1997; Williams 1997).

London has long been regarded as a sporting capital, but this reputation has been reinforced in recent years. The London 2012 Olympics encouraged the construction of new sporting venues in East London, while in other non-central areas a wave of redevelopment has embedded tourism offering within existing sports spaces. Therefore, in line with one of the central themes in this book, this chapter also investigates the ways in which sports stadiums have expanded the spatial reach of tourism and (re)distributed visitors outside central London. Such facilities may add to the appeal of the locality, encouraging visitors to engage with neighbourhoods outside of the popular central tourist districts.

London's Sport Tourism Venues

Since the 2012 Olympics Visit London has extended their promotion of the sporting attributes of London. The variety of world-class sporting facilities and the expertise to organise sporting events encouraged the city to stage numerous world championships in the years following the Olympics (Greater London Authority 2014). London has consistently topped the Global Sports Cities Index, which assesses more than 700 multi-sport events over a rolling period

(7 years past to 7 years forward) to compare those cities active as hosts of major sporting events (Sportcal 2017).

In recent years London has seen the development of many temporary sporting facilities which can serve local and the domestic and international tourist markets. For example, winter sees the creation of many ice-skating rinks, located both at popular tourist sites such as the Natural History Museum and Tower of London as well as in locations outside of the central area (such as Tobacco Dock in Wapping and Canada Square Park in Canary Wharf). During spring and summer, the closure of city streets provides spaces for sporting opportunities such as the London Marathon and Ride London, events that provide opportunities for professional and amateur athletes to compete on traffic-free streets.

These events, while perhaps restricting regular community use of local infrastructure, add to the sporting appeal and assets of London, alongside the permanent sporting venues. Across London, arenas and stadiums are the home for rugby, football, and cricket teams as well as providing spaces to watch a variety of sports including athletics, cycling, basketball, and field and ice hockey. There are also entertainment facilities which accommodate sporting events. For example, the O2 arena has played host to the ATP Tennis Finals since 2009 and a regular season NBA Basketball game since 2011. Alexandra Palace is the host of the PDC World Darts Championships and the UK Masters Snooker tournament.

Notwithstanding the numerous entertainment spaces that play host to sporting events in the city, Table 5.1 exhibits the key sporting infrastructure available in the city. All these venues have the potential to host sports tourism and enhance the image of London, but this chapter predominantly focuses on the stadiums that currently offer attractions for sports tourists through the inclusion of tours and/or museums (Group 1). The chapter also includes discussion of the London Stadium, one of the sporting venues in the Queen Elizabeth Olympic Park (Group 2). While the venues listed in group 3 are not specifically discussed in this paper Crystal Palace National Sports Centre is worthy of comment here. This facility offers limited touristic appeal, despite being the national centre for athletics. Built in the 1960s, many years of underinvestment caused management and operational issues, compounded further since 2012 by the move of some major athletics meetings to the London Stadium. In 2018, the Mayor of London initiated a review of the Centre to consider design and development options in an attempt to achieve a long-term future for the location (Majendie 2018). Thus, at a time when many London facilities are thriving, this stadium is facing greater competition and an uncertain future.

Many stadiums are the home of a sports team and so the image projection of London as a sports destination is achieved in part by the activities of the sports team and its regular use of the stadium. Table 5.1 emphasises that football (soccer) dominates London's sportscape. This is driven by football's significance as the national sport, which imbues a sense of cultural authenticity. The

Group 1: Stadiums with Tours/Museums
Craven Cottage (home to Fulham Football Club)
Emirates Stadium (home to Arsenal Football Club)
Kia Oval (home to Surrey Cricket Club)
Loftus Road (home to Queens Park Rangers Football Club)
Lords (home to Middlesex County Cricket Club)
Stamford Bridge (home to Chelsea Football Club)
The Den (home to Millwall Football Club)
Twickenham (home to England Rugby Union)
Wembley Stadium (home to the England national football team)
White Hart Lane (home to Tottenham Hotspur Football Club)
Wimbledon Tennis (home to the All England Lawn Tennis and Croquet Club)
Group 2: Sports Venues constructed for London 2012
London Stadium (originally the Olympic Stadium and currently home to West Ham United Football Club)
Copper Box Arena
Lee Valley Velo Park
London Aquatics Centre
Eton Manor (Lee Valley Hockey and Tennis Centre)
Group 3: Other Stadiums/Sports Grounds
Allianz Park (home to Saracens, a rugby union club)
Brisbane Road (home to Leyton Orient Football Club)
Crystal Palace National Sports Centre (hosting national athletics meetings)
Queens Club (private sporting club hosting a major tennis event)
Selhurst Park (home to Crystal Palace Football Club)
The Stoop (home to Harlequins, a rugby union club)
The Valley (home to Charlton Athletic Football Club)
Trailfinders Sports Ground (home to London Broncos rugby league club)

Table 5.1: London Sports Stadiums and Venues. Source: Devised by the Author.

touristic appeal of football is also influenced by the international draw of the English Premier League (EPL) with research revealing that 2 per cent of visitors to London are likely to watch live football during their stay (Visit Britain 2015). The media rights value of the EPL is more than double its nearest rival, Spain's LaLiga (Sport Business 2016), reflecting its appeal domestically and internationally – it is broadcast to 156 countries with estimated audiences of

Figure 5.1: The New Warner Stand at Lord's Cricket Ground (Photo: Andrew Smith).

4.2 billion (Eurosport 2015). Consequently, many domestic and international tourists want to attend a live match or visit the stadium of EPL teams when they come to London.

For Premier League football clubs, the financial return from the operation of stadium tours is often small in relation to the revenue earned from broadcasting rights and football player trading. Furthermore, while other sports receive less from broadcasting rights, financial accounts (SCC 2016; Rugby Football Foundation 2015) suggest that the income from stadium tours and museums is usually only a small percentage of the total revenue earned by these businesses. Revenue from sponsorship, ticketing, hospitality and corporate events substantially overshadow earnings from tourism. However, offering access to the stadium via tours supports the brand and delivers an important fan experience.

After the London 2012 Olympics the venues created in the Queen Elizabeth Olympic Park (henceforth termed the Olympic Park) have been altered, particularly with regards to the London Stadiums and the adaptations required to make it the new home of West Ham United Football Club. The stadium now offers tours, which acknowledge its role as an athletics stadium as well as an EPL ground. As at the Emirates Stadium (the home of Arsenal Football Club in north London), tours are predominately operated as self-guided (using multi-language audio guides). Other London stadiums offering tours use human guides to direct and inform visitors. Several stadiums, including Emirates Stadium, offer a premium tour that employs a well-known ex-player to accompany the tour group to further enhance the visitor experience.

Research Method

This chapter draws on two sources of data. Firstly, a mystery shopper exercise was completed at eight of the London stadiums. Ethical debates regarding participant observation in the form of mystery shopping largely focus on issues of privacy and informed consent (Oliver and Eales 2008) but are countered by assertions that information is effectively in the public domain, available to anyone who seeks the information or experience (Jorgensen 1989, Ng Kwet Shing and Spence 2002). Consequently mystery shopping is seen as a 'quite mild and on the face of it harmless form of deception' (Hammond and Wellington 2013, 61) and therefore as a research approach has largely become mainstream (Hudson et al. 2001). Although mystery shopping may be perceived as deceptive it is used extensively by commercial organisations to assess service delivery and performance (Wilson 1998). While there has been discussion of the ethical use of mystery shopping for critical appraisal of human or competitor performance (Ng Kwet Shing and Spence 2002), in this case it was undertaken to gain appreciation of the norms of tour design and delivery. This insight informed the critical analysis of the main data set drawn from Trip Advisor reviews. Gaining experience as a participant allowed enhanced interpretation of these reviews written following other people's experiences of stadium tours.

The main data set that underpins this chapter consists of more than 7,000 reviews posted on Trip Advisor, which detail the visitor experiences of stadium tours. Table 5.2 summarises the number of Trip Advisor reviews for each venue analysed.

Trip Advisor relies on user-generated content to provide travel-related reviews which may act as a form of word-of mouth recommendation to influence the decision-making of others (Gretzel and Yoo 2008). The scale of such resources (more than 535 million reviews have been posted on Trip Advisor) provides insights from a variety of users. Thus, it is possible to achieve 'insight extraction' (Gandomi and Haider 2015, 140) through analysis of such data.

The dataset was collated by downloading the full Trip Advisor review for each venue. In some cases, there are separate Trip Advisor pages for the venue tours and the match-day experiences. However, in other cases there is no separation of reviews; thus the data set included only those reviews which made specific mention to the word 'tour' (noted in Table 5.2). In total 7040 reviews were collected for analysis.

Nvivo was used to analyse the dataset, a qualitative software tool which helped to manage the thematic analysis of each review. Thematic analysis identifies ideas from the data (Guest et al. 2011), allocating codes (or names) which represent these so that other examples can be identified and compared. It also allows contrasting cases to be recognised and evaluated (Flick 2009). This allowed advanced levels of abstraction (Punch 2005) to ensure the development of overarching conceptual ideas from the dataset. Analysis of the data revealed a wide variety of issues linked to the experience of stadium tours. However,

Stadiums with Tours/Museums	Reviews (November 2017)
Craven Cottage (home to Fulham Football Club)	293 (10 mentioning 'tour')
Emirates Stadium (home to Arsenal Football Club)*	1,799 (tour specific site)
Kia Oval (home to Surrey Cricket Club)*	189 (8 mentioning 'tour')
Loftus Road (home to Queens Park Rangers Football Club)	117 (4 mentioning 'tour')
London Stadium (home to West Ham United Football Club)	1,419 (91 mentioning 'tour')
Lords (home to Middlesex County Cricket Club)*	1,275 (657 mentioning 'tour)
Stamford Bridge (home to Chelsea Football Club)*	1,602 (tour specific site)
The Den (home to Millwall Football Club)	96 (80 mentioning 'tour')
Twickenham (home to England Rugby Union)*	1,381 (122 mentioning 'tour')
Wembley Stadium (home to the England national football team)*	4,816 (1,000 mentioning 'tour')
White Hart Lane (home to Tottenham Hotspur Football Club)*	395 (tour specific site)
Wimbledon Tennis (home to the All England Lawn Tennis and Croquet Club)*	998 (museum) 1,287 (274 mentioning 'tour')
Total Reviews	15,667 (7,040 = museums/tours)

Table 5.2: Trip Advisor Reviews.
Visited as part of Mystery Shopper research.

the findings presented in this chapter focus only on the visitor experience of London sports stadiums as tourist attractions.

The Appeal of Sports Facilities as Tourist Attractions

Sports venues act as the home base for a sporting team and they can be venerated for such associations. These modern-day shrines are seen as sacrosanct spaces that appeal to fans but also to wider audiences. Substantive research into the 'Fan' has been developed over the years, focusing particularly on fan identification (Sutton et al. 1997). Shortened from the term fanatic, the fan is seen to hold obsessive, devotional emotions, akin to a religious or spiritual fervour. In a sporting context being a fan suggests a taking of sides, with affiliation to a group with similar beliefs. Fandom suggests perseverance, a long-term relationship and a level of self-identification with the sport (Jones 2000). In recent years academic discussion of the nature of fandom has recognised changes in sporting affiliation, heavily influenced by media coverage (Williams 2007).

Being a fan can stimulate continued engagement even in cases where the cost of participation in a leisure activity is seen to be greater than the benefits experienced. In other words, there is an irrational element of fan consumption that defies established models of consumer behaviour. Fandom occurs at many levels, with supporters of the team or the sport attracted to those stadiums or arenas that act as spaces where the sport is displayed.

There has long been significant research linking sport fans to tourism. Fifteen years ago, Gibson (2003) explored the idea of the 'fan as tourist' with an investigation of the supporters of the University of Florida American football team. She concluded that leveraging tourism benefits from visiting fans is possible. Furthermore, Sutton et al. (1997) drawing on the work of Wann and Branscombe (1993), acknowledges three types of fan (social, focused, vested) each holding differing levels of identification with their teams. Consequently, it isn't just the ardent, vested fans that visit sports spaces. Those with a love of sport generally (with only a relatively passive, social interest in a specific team) may also want to visit. Furthermore, tourists travelling with family and friends are likely to visit these sites together; thus non-fans also engage with such experiences (Table 5.3).

The reviews show that the tour experience is often unexpectedly positive for non-fans. Comments intimate that reviewers are 'NOT a fan BUT ...', suggesting that tours of sports stadiums have a wider appeal. Some literature considers the degree to which someone is a fan of a sport or team (Wann and Branscombe 1970, Hunt et al. 1999), but there is little empirical research considering engagement by non-fans. If fandom comprises a degree of commitment and loyalty to a sport or club (Pedro, Carmo, and Luiz 2008) then, appreciating that, those who do not consider themselves fans but who are still engaging in activities akin to low levels of fandom are exhibiting contradictory behaviours. Gray (2003) recognises this in his discussion of anti-fans (those with an informed distaste) and the non-fan who engages with experiences but with lower levels of involvement. This research suggests that engagement with the tours is often initiated for the benefit of a sports fan but participation reaches further to offer non-fans (those with low engagement with the sport or club) something of interest. Consequently, market size can be expanded to include a wider group of tourists. This helps to explain why sports venues – previously considered to have niche appeal – are now becoming mainstream attractions.

This research also reveals that the association of sporting places as spiritual or sacred spaces is clearly established in the mind of London visitors. Brooker (2017, 157) recognises that fans will 'travel across the world to often mundane places that fandom has made sacred' including the football stadium (Jorgenson 1995). In evaluating Trip Advisor reviews of stadium tours there are numerous contributors that liken their visits to pilgrimages (Joseph 2011, Gammon 2004) with the sports spaces honoured as holy or 'hallowed' (Table 5.4).

'Great Tour' I am not a true fan of Association Football also known in the USA by the original British nickname-soccer, but I greatly enjoyed this tour. Our guide was very entertaining, and I learned a lot about the business models used to fund this multi-billion dollar industry.	'Hate football but loved this tour!' I have no interest in football, or footballers, never mind stadiums, but my children do (even though they are not Arsenal fans) so I came along with low expectations. I was very happy with the tour and think if you are ever going to see round a stadium, this is a good one to see.
'Interesting' My boyfriend got 2 tickets for Christmas, I'm not a huge football fan so wasn't too bothered with this but actually quite enjoyed it.	'Amazing, boyfriend was chuffed!' Booked to do the tour for my boyfriend's birthday, I'm not really into football but I really enjoyed it.
'Not Chelsea fans and still enjoyed this tour and museum!' Museum really good and this kept us all entertained, the tour was good but felt could be better with footage of some past press conferences.	'Very cool stadium tour'‡ I went there with a bunch of friends to take the stadium tour and I think all of us genuinely enjoyed learning about the stadium itself and West Ham Football Club, even people with little interest for football. The videos/audio and the tour guide made it extra special and really enjoyable.
'Interesting tour' Went on a weekday morning for a 90 min tour. The tour takes you all around the grounds - specifically to centre court, media room and other courts. Even though I am not a tennis fan, had an interesting morning at this place.	'Loved the stadium tour' As a QPR fan for most of my life, I was always going to love the Loftus Road Stadium Tour. However, as I had a spare ticket, I dragged my sister along who is not a football fan at all and she really enjoyed it too.

Table 5.3: Non-fan Engagement with Stadium Tours.

These sports grounds are also seen as spaces of worship. For example, Lord's cricket ground is referenced as 'A worship place for cricket lovers', 'The Vatican of cricket' and 'The Mecca of cricket'. The terminology reflects the quasi-religious appeal of such attractions. Lord's is not unique in this association; for example, Twickenham is described as 'The temple of Rugby'. Bale (1995) acknowledges that the stadium is a cathedral for the masses while Gebauer (2010) further adds that the architecture of the stadium, with sacrosanct ground at its centre, allows the separation of the profane fan from the saint-like players. Price (2001, 3) suggests that sports fans 'exhibit a kind of devotion that is often described in terms of religious dedication or intensity'. Consequently, stadiums provide the space for adoration of players by sports tourists regardless of the presence of sporting action.

'Dream come true!' A fantastic experience for our 'true Blue' son and the rest of the family. From sitting in Jose's seat to exploring the dressing rooms, nothing was off-limits except the <u>hallowed turf</u>! (Chelsea Tour)	'Take the ground tour!' We were at The Oval for an AGM. After the meeting we were taken on the Ground Tour. We were allowed onto the <u>hallowed turf</u> and taken over to the 'square' to see how the ground looks to the players. (Oval Tour)
'The Home of Rugby' Always a pleasure to enter the <u>hallowed halls</u> and sit in the greatest stadium in rugby. We have been many times and will continue to do so. (Twickenham Tour)	'Visit Wimbledon if you are a tennis fan' Very well worth visit to this <u>hallowed ground</u> of tennis. The tour is good and gives good info on how the tournament works. (Wimbledon Tour)
'Stand on the <u>Hallowed turf</u>!!' Access all areas of interest that you wouldn't usually see and touch the FA cup. (Wembley Tour)	'The Home of Cricket' Astounding place to visit. There's a visit to the pavilion, a walk down the Long Room and a trip to the <u>hallowed turf</u>. (Lords Tour)

Table 5.4: Hallowed Space.

For many holy sites that have touristic appeal the balance between times for worship and times for visitation must be managed (Brown et al. 2009) and for sport stadiums these distinctions can be clearly identified as days of worship are determined by match schedules. Visiting the stadium at times of worship (during matches or competitions) may seem to be the preserve of the vested ardent fan but the commercialisation of sport has limited access to such opportunities. The growth of corporate entertainment packages within sports stadiums introduces 'a less passionate fan base famously derided by Manchester United captain Roy Keane' for its lack of knowledge of the sport (Slack and Amis 2004, 270), which excludes other fans. Thus, tours provide access to those otherwise unable to visit such venerated spaces during the times of scheduled 'worship'.

In London the international appeal of the EPL plays an important role in promoting London stadiums. Arsenal and Chelsea prove the most popular teams to watch (Visit Britain 2015) but lesser-visited teams such as Fulham also offer appeal for some tourists (Table 5.5).

Just as Arsenal and Chelsea dominate the live match market so they also dominate the stadium tours/museums market, both receiving in excess of 200,000 visitors annually. This is almost double the number of tour visitors to Wimbledon AELTC and Twickenham combined (86,000 and 30,000 visitors respectively) (Visit England 2016).

Team	Number of International Tourists to Live Matches Annually
Arsenal; Emirates	109,000
Chelsea; Stamford Bridge	89,000
Wembley Stadium	51,000
Tottenham Hotspur; White Hart Lane	40,000
Fulham; Craven Cottage	30,000

Table 5.5: International Tourists Attending Football Matches. Source: Visit Britain 2015.

Figure 5.2: Poster on the London Underground Advertising Tours of Arsenal Football Club (Photo: Andrew Smith).

Stadiums as Tourist Infrastructure

Sports stadiums are important parts of the urban infrastructure of London, with the commercial power of the EPL driving waves of investment in the redevelopment of stadiums home to Premier League teams. At the time of writing (2017/18 football season) Tottenham Hotspur are playing their home matches at Wembley Stadium while their ground at White Hart Lane is being rebuilt.

Figure 5.3: Wembley Stadium on FA Cup Final Weekend (Photo: Andrew Smith).

The lack of a home ground has not put paid to Tottenham Hotspur stadium tours, with specialist 'Spurs at Wembley' tours offered on the days adjacent to their home matches. Furthermore, the importance of serving the visitor market is revealed as the club has already announced that tours will be available at the new stadium when completed. The new stadium has been designed to incorporate a permanent visitor centre, housing a museum and Hall of Fame.

Redevelopment of Stamford Bridge is also planned by Chelsea FC, with expectations that this will increase match-day capacity by about 50 per cent. Unsurprisingly, given demand levels at the existing stadium, the proposals include a space for tours and a museum. The extent to which touristic infrastructure is embedded into redevelopment plans for stadiums is evident, and stadiums incorporating hotels (such as Twickenham), museums and restaurants can encourage tourists to increase their dwell-time in the local area. The development of sports stadiums that function as entertainment zones with touristic appeal has featured prominently in the redevelopment of urban areas (Hinch and Higham 2011). Historically, stadiums served a largely local market but, today, larger stadiums, increased car ownership and wider geographical spread of fans means stadiums are designed to cater for regional, national and even international audiences.

Sports-related development policies are often justified for their trickle-down benefits to communities (Jones 2001, Stevens and Wootton 1997) and stadium redevelopment has been common in London in recent decades. Completed stadium-led regeneration schemes include Wembley Stadium (rebuilt between 2002 and 2007, see Figure 5.3) and Arsenal FC's move from its old ground in Highbury to the Emirates stadium in 2006 (London Assembly 2015). Both grounds included aspects of regeneration for the surrounding districts. The success of such redevelopment is debated (Davies 2005, Bourke 2015), particularly considering the impacts to local communities as well as to supporters and other visitors. Collins (2008) argues that stadiums are often unsuccessful as development catalysts because they result in unevenly distributed benefits, leading to social and spatial inequalities. There are also concerns that benefits may only be ephemeral, and with redevelopment plans introducing infrastructure that is likely to be in existence for many decades it is challenging to provide a long-term cost-benefit appraisal. Despite these reservations, clubs continue to drive redevelopment plans led by an enthusiasm to gain increased revenue from ticket sales, corporate hospitality, stadium naming rights, sponsorship and non-match-day rental earnings (Zinganel 2010).

Recently constructed stadiums have incorporated tourist facilities within the building from the outset while older stadiums are trying to incorporate tours and museums into existing facilities. Tours vary in complexity and popularity with Arsenal Emirates and the London Stadium offering self-guided tours while others use trained individuals to guide visitors through the building. In some cases, the individuals are volunteers (Twickenham and Kia Oval) while Wimbledon AELTC uses accredited guides in an attempt to ensure a quality experience. In all cases the tours provide backstage access to spaces which are not seen by match-day attendees. This is an important part of the offering, sought after by visitors (Table 5.6). Gaining access to spaces usually reserved for the sports players is valued.

The reviews show that tourists value the opportunity to move beyond the public spaces of sports infrastructure. The opportunity to sit in the manager's chair, see the players' changing rooms or visit the press areas (including those spaces where players are interviewed for TV) provides the tourist with meaningful engagement with the physical environment of such buildings. The power of the backstage to generate feelings of adoration and veneration (Gammon and Fear 2005) is evident in the data for London stadiums. Expanding on the work of Goffman (1959), the importance of the backstage was recognised by MacCannell (1973) and has since been extensively examined in tourism research (Pearce and Moscardo 1986, Cohen 1979, Sharpley 2008) including more recent discussion of access to the backstage when venerating sports sites and people (Hinch and Higham 2005).

Accessing spaces usually only available to the elite few appears important for fans who relish the opportunity to 'cross the symbolic boundaries that distinguish the world for the audience and the worlds of the performer or privileged' (Ramshaw et al. 2013: 19). The creation of an organised stadium tour converts the backstage to a frontstage, as access is no longer restricted to the

'Great tour - whether you're a fan or not!' This new tour in the new West Ham stadium is a great experience - whether you're a fan or not! It's fascinating to see backstage, including the VIP sections, dressing rooms and dugout.	'Good fun!' Of course, Wembley needs no introduction, and getting to go backstage - quite literally - is a great experience. The tour takes you through the changing rooms, pitchside area, press conference room, royal box and more.
'Birthday boy's dream day out' We started with a good look at the stadium which if you have never seen one is a vast and impressive structure plus a tour of the press room in a chance to sit in the manager's chair. Then an insider view of the huge spa like changing rooms and then a lifelike exit through the dug out onto the pitch to the sound of applause. You can't fail to feel the energy and adrenalin which the players must experience. It's a backstage pass into a footballer's life.	'Pretty cool backstage look!' The Arsenal stadium (audio) tour is pretty cool! It gives a very detailed look behind the scenes. It really takes you to places which you normally can't access. Locker rooms, business club, the pitch. Truly awesome experience.
'Stadium Tour Fulham' Treated my husband who is a staunch Fulham fan to a trip around the ground. It was very informative and you went behind the 'scenes' to where the players change and where they wives wait while they are playing.	'Fascinating behind the scenes view' Excellent guide - obviously enthusiastic. Loved seeing the Long Room and the surprisingly sparse dressing rooms. Also enjoyed the amazing view from the state-of-the-art media centre. (Lords Cricket Tour)

Table 5.6: Touring the Backstage.

few. MacCannell (1976) highlighted the existence of different types of backstage, including some altered to be accessible to tourists. Regardless of adaptations, tours offer access to spaces within the stadium that are endowed with a special status (Gammon 2011). Furthermore tours provide the fan with a sense of intimacy with the players and/or teams through greater appreciation of the spaces they inhabit (Ramshaw and Gammon 2010).

Motivations to take tours of sports venues and their associated museums are driven in part by a desire for a nostalgic engagement with the hidden aspects of stadiums. Consequently sport tours, halls of fame and sports museums are 'a unique opportunity for devoted pilgrims to enter areas that are otherwise restricted, providing authentic insight and an otherwise unforgettable backstage experience' (Wright 2012: 197). Such experiences are proliferating (Kellett 2007) to a point that they are embedded into stadium design and promotion.

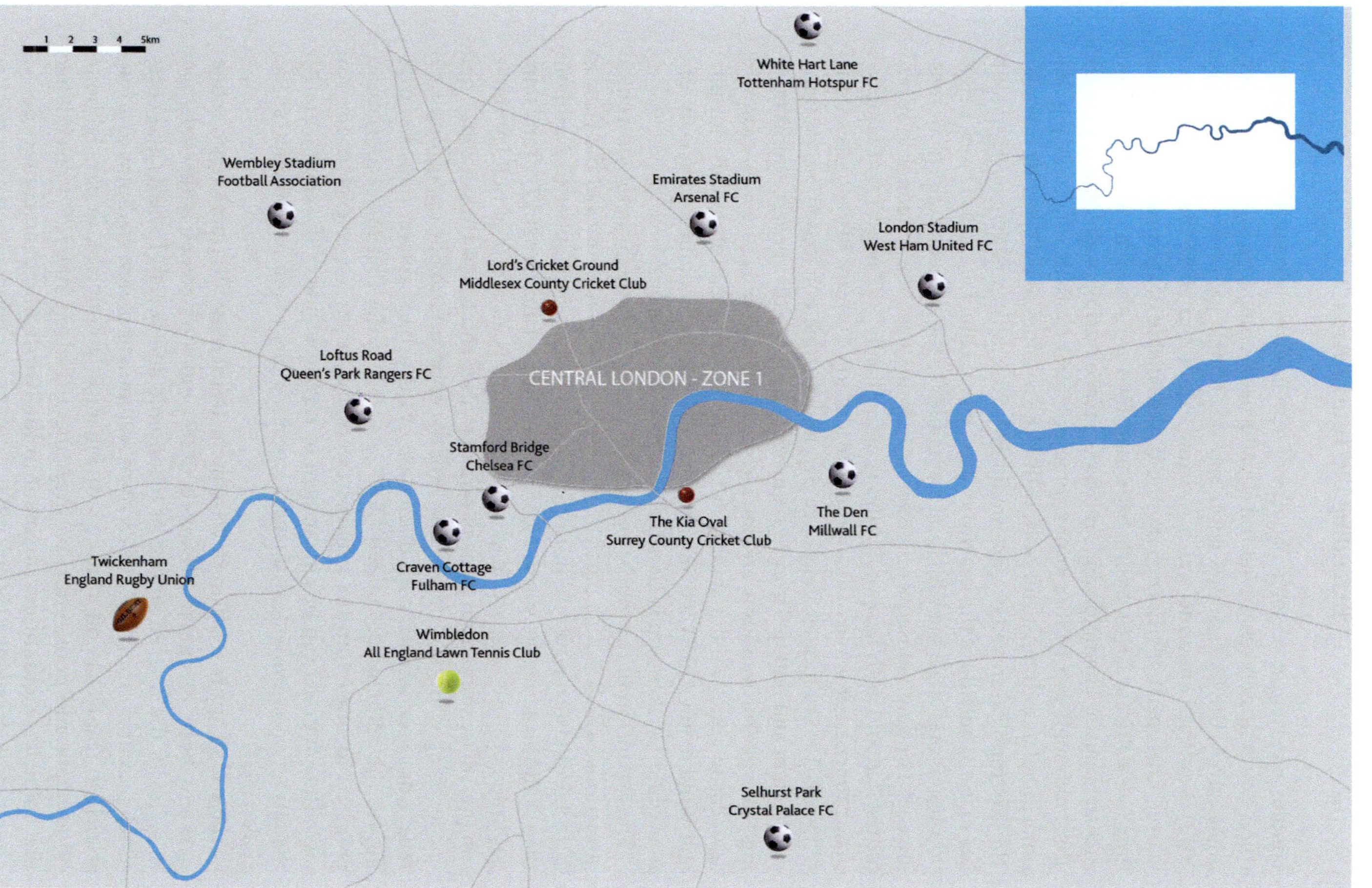

Figure 5.4: Non-Central Locations of London's Key Stadiums (© Mason Edwards).

Extending the Tourist Area

London experiences seasonal patterns of demand which are more pronounced when examining international travel patterns as compared to domestic overnight travel. The latter shows fewer signs of imbalance across the year (Kyte 2012). Visiting London to watch sporting events can affect the travel season, with football particularly encouraging travel outside of the peak UK summer time (Visit Britain 2015). Sports have their own seasons that can mean events such as finals or championships must take place during peak holiday periods. Thus, although sport tourism has some potential to assist in addressing seasonal imbalances it must still work within the confines of the sporting calendar. The geographical spread of sport stadiums across the city has encouraged tourists to move outside of the heavily visited central core. Discovering new neighbourhoods and engaging with local teams adds to the authentic appeal for these areas of London. Consequently, sports stadiums have the potential to spread tourist demand both temporally and geographically. The majority of London's major tourist attractions (including 9 of the top 10 most visited attractions) (ALVA 2017) are located in an area denoted as zone one (based on the transport system and outlined grey in Figure 5.4) while all the sport stadiums offering tours and/or museums are outside of this area (see Figure 5.4). Many are adjacent in zone two but stations for White Hart Lane (Tottenham Hotspur) and Wimbledon AELTC are in zone three, Wembley Stadium station is in zone four while Twickenham is classified as a zone five station.

The expansion of the tourist zone is encouraged through these non-central attractions, which can be particularly important at busy times of the year. In 2016 London received one-third more visits in July-September (5.2 million) than it did in January–March (3.87 million) (Visit Britain 2017). Thus tourists' intensive use of space, particularly during peak seasonal holiday times, can pressurise the central parts of popular cities and towns (Russo 2002), and encouraging visitors to move outside of the core can provide opportunities to relieve this congestion. In his discussion of the urban explorer tourist searching for the authentic, Maitland (2017, 67) recognises that many tourists are prepared to move into the suburbs to enhance their 'experience of the real city'. Thus demand can be encouraged to travel outside of the central zone if the attraction is evident, with sport stadiums often seen as authentic spaces (Bale 1993).

The use of sport tourism development to increase destination attractiveness has been recognised (Daniels 2007) but this comes with challenges in terms of the uneven spread of (economic and social) benefits and costs. Much literature from the USA on the development of new stadiums to attract franchise teams has often been critical, raising concerns over neighbourhood blight and poor return on investment (Nelson 2001). However, issues that influence stadium development in the North American context do not always exist elsewhere (Thornley 2002) and, for London, many of the existing

'If you haven't been to the stadium for an event it's interesting.' The major focus is now on the West Ham football team but there is still a lot of Olympic & para Olympic detail. It is a self-guided audio tour but there are many helpful staff along the way. As well as a view of the arena you get to see the main hospitality area, the home dressing room, the indoor warm up track, the tunnel and dug-out. At the end you can return to the club shop to get a souvenir certificate of your visit.	'At your own pace' We did the tour on a non-match day and were pleasantly surprised. We had a warm welcome, were given our head-sets, tablet and a quick introduction to the tour, then off we went following the signs and playing the appropriate video on our tablets. Really interesting, such as when we walked through the tunnel a guide was on hand to chat about how they remove the seats for athletics. When you have finished you receive a voucher to take back to the club shop to get a per-sonalised certificate of your tour (either in West Ham or Olympics format) and you receive 15 per cent off in the cafe. Well thought out and very pleasant experience.
'Proud to be a Hammer - Enjoyed the Tour' Loved our trip to see our new ground - London Stadium. Will miss Upton Park after going for 40 years, but enjoyed the look around the stadium. Can spend as much time as you wish in each section, pitch, tunnel, changing rooms, etc.	'Great Experience!!' I am not a super football fan but I must say that the tour of the stadium was great and very interesting... plus was my first time in a completely empty Olympic stadium... it literally left me breathless! I would totally recommend it!
'Stadium Tour' This is more of a West Ham tour that of the Olympics – there is limited Olympic stuff to see or hear about. We enjoyed that it is self- guided audio tour so you can take your time and explore at your own pace. You get to go into the changing rooms, warm up track and on to the pitch dug out as well as sitting in the exclusive seats. The views are great and there are assistants on hand to ask questions of.	'Not just a football stadium' I loved the stadium! It brought up memories of the glorious Olympic Games in 2012. The tour was very comprehensive and not just focused on football which I appreciated. It was very interesting to hear how it has changed since it was built. The audioguide in the tour is really good, it has a lot of quality videos and information.

Table 5.7: London Stadium Tours.

sports stadiums have been long established in their neighbourhoods. There-fore, notwithstanding the recent waves of redevelopment, these places are rooted in their local areas and thus offer tourists a means of engaging with an authentic local space. One key exception to this is the development of the Olympic Park stadiums, where the construction of new stadiums occurred alongside the construction of new residential property, shopping and other local amenities and services.

The need for regeneration of the Olympic Park area of East London was evident, with much of the site containing polluted waterways and brownfield spaces. The construction of permanent and temporary stadiums was seen as an opportunity to transform a run-down part of East London and during the construction phase policy rhetoric focused on the legacy of Olympic-led regeneration. In the period since the Games, the Olympic Park has been redesigned to capture the legacy benefits of the infrastructure (Latuf de Oliveira Sanchez and Essex 2017). The success of this development is hard to assess, however, as plans for the infrastructure have changed since the initial inception phase (Azzali 2017). This is particularly evident with the main Olympic Stadium. Originally designed to be reduced in size (to 25,000 seats) and be used as an athletics stadium, it is now a multi-sport arena with capacity for 66,000. Consequently, it is now capable of hosting EPL football matches and is the home ground of West Ham United. This has provided the stadium with a regular schedule of events as well as increased media coverage through televised matches.

Redevelopment of the Olympic Park stimulated service sector growth in the area, providing new spaces for the consumption of leisure. Estimates suggest there have been more than 15 million visits to the venues in the park, with one million spectators heading to the London Stadium for music and sports events, including West Ham United home games (LLDC 2017). Converting the main athletics stadium into a multi-use arena embedded facilities to operate self-guided tours. The design of the tour needed to acknowledge both the Olympic history of the stadium as well as its status as the new home of West Ham United Football Club. Reviews suggest that tours have, for the most part, been successful in achieving an appropriate balance (Table 5.7). The reviews highlight that, in line with other stadiums in London, the tour offers access to spaces not usually accessible to the public during sporting events, while the use of multimedia technology within the tour has also helped maximise the visitor experience.

The sharing of stadiums means separate fan bases may be attracted to the location. Some European football teams in the same league share grounds (for example AC Milan/Inter Milan and AS Roma/SS Lazio) and cross-sport sharing occurs frequently between football and rugby teams. While the London Stadium is not currently shared by two teams, its recent history as an Olympic stadium still provides an important draw to some tourists interested in the Olympics. Furthermore, club relocation for West Ham United also has an impact on the experiences of visitors. Relocation to a new ground is not a recent phenomenon in football (Vamplew et al. 1998, Horak 1995); thus fan allegiance to place can transition when teams move to a new home (Brown 2010). This is controversial (Maguire and Possamai 2005) but there have been successes when support has come from the fan base (Vamplew et al. 1998, Tallentire 2018). Thus, in such cases demand for tours can be stimulated by a desire to see the new home of a team.

Figure 5.5: The London Stadium in Athletics Mode (Photo: Andrew Smith).

While the Olympic Park hosts tours of the Aquatics Centre, Velodrome and London Stadium, it is the last of these that attracts the greatest demand. From the reviews it is clear that its Olympic history is still an important part of its appeal. Many Olympic cities mark anniversaries of the Games (Cashman 1999) and London is no exception, with the stadium hosting an athletics meeting every July. Consequently, annual TV coverage displaying the stadium as more than an EPL ground continues to drive diversity in its appeal to visitors. Sports stadiums as part of urban regeneration programs have often relied on TV coverage (of the variety of events taking place at the stadium) to stimulate awareness and encourage international visitation (John et al. 2013). Media coverage can create place recognition in the mind of visitors. This can, in turn, encourage tourists to move outside of the central tourist districts into other areas of the city.

Television coverage of mega-events such as the London 2012 Olympics and annual events such as the Wimbledon tennis championships can enhance national and international awareness of these sports facilities; consequently 'because of its global reach, telecasting plays an active role in defining, shaping and changing national images around the world' (Zeng et al. 2011, 41). Events as a component of the destination's product can leverage media attention to promote the place image (Brown et al. 2004), hence the construction of iconic buildings and the staging of events being strategically employed to assist in the re-imaging of cities (Smith 2005). Thus sports stadiums are thought to enhance perceptions of the destination and add marketing appeal (Thornley 2002).

Conclusions

London's rating as the 'world's best sporting city' (London and Partners 2018) has been driven by the success of hosting peripatetic events such as the London 2012 games, the 2015 Rugby World Cup and the 2017 IAAF athletics world championships alongside the long established sporting schedule that includes the Wimbledon tennis tournament and EPL football matches. As well as enhancing destination image, sporting events were found to contribute £1.67 billion to the London economy between 2013 and 2016 (London and Partners 2016). London's growing reputation as a sporting city has been part of the appeal for the NFL to use it as a base for its international series of games, from one a year in 2007 to four a year in 2017. TV coverage of such events further extends awareness to potential visitors, as views inside stadiums and around the historic city promote the touristic offering.

Extending the tourist offering outside of the central London core is valuable in managing demand pressures as well as offering tourists the opportunity to engage with different localities and neighbourhoods. The provision of scheduled tours and associated team or sport-related museums has made sports stadiums in London a robust part of the touristic offering. Accessibility, vitally important when stadiums are used for sporting events, ensures transport links exist for those tourists coming on non-match days. Construction of new stadiums across London has led to the expansion of the tourist region and, with the new stadium at White Hart Lane due for completion for the 2018/19 football season, the number of sport tourism attractions is set to further increase.

Sports stadiums across London have attracted many visitors to watch games, take backstage tours and, in some cases, to use the facilities to participate in sporting activities. The appeal of sport as a primary or secondary motivator brings tourists to the city both during the peak holiday season and during less popular times, bolstered by the appeal of sporting events. Reviews of the visits to sports stadiums are overwhelmingly positive, for fans and non-fans alike. Despite receiving only about one-tenth of the numbers that visit attractions such as Shakespeare's Globe or St Paul's Cathedral, the scale of visitation to London sports stadiums (for tours and live sports events) reveals that such infrastructure is an important component of the London tourist product and helps extend tourism beyond central areas.

References

ALVA. 2017. Visits made in 2016 to Visitor Attractions in Membership with ALVA. Association of Leading Visitor Attractions. Available at http://www.alva.org.uk/details.cfm?p=607 (accessed 18 January).

Azzali, Simona. 2017. 'Queen Elizabeth Olympic Park: An Assessment of the 2012 London Games Legacies'. *City, Territory and Architecture*, 4(1),11.

Bale, John. 1993. *Sport, Space, and the City*. London: Routledge.

Bale, John. 1995. 'Introduction'. In John Bale and Olof Moen, eds, *The Stadium and the City*. Keele: Keele University Press.

Bourke, Joanna. 2015. 'Are Shiny New Football Stadiums a Game-changer for Communities?'. *The Independent*, 23 December. Available at http://www.independent.co.uk/news/business/analysis-and-features/are-shiny-new-football-stadiums-a-game-changer-for-communities-a6785031.html (accessed 16 January 2018).

Brooker, Will. 2017. 'A Sort of Homecoming: Fan Viewing and Symbolic Pilgrimage'. In Jonathan Gray, Cornel Sandvos and C. Lee Harrington, eds, *Fandom: Identities and Communities in a Mediated World*. New York, NY: New York University Press.

Brown, Adam. 2010. 'Come Home'. In Sybille Frank and Silke Steets, eds, *Stadium Worlds: Football, Space and the Built Environment*. Abingdon: Routledge.

Brown, Graham et al. 2004. 'Developing Brand Australia: Examining the Role of Events'. In Nigel Morgan, Annette Pritchard and Roger Pride, eds, *Destination Branding: Creating the Unique Destination Proposition*. Oxford: Elsevier Butterworth-Heinemann.

Brown, Jan, Phillips, John, and Vishwas Maheshwari. 2009. 'Consumption, Sacred Places and Spaces in Profane Contexts: A Comparision Between the UK and India'. In Steve Brie, Jenny Daggers and David Torevell, eds, *Sacred Space: Interdisciplinary Perspectives within Contemporary Contexts*. Cambridge: Cambridge Scholars Publisher.

Cashman, Richard. 1999. 'Olympic Legacy in an Olympic City: Monuments, Museums and Memory'. In Richard Cashman and Anthony Hughes, eds,

Staging the Olympics. The Event and its Impact. Sydney: University of New South Wales.

Charney, Igal. 2007. The Politics of Design: Architecture, Tall Buildings and the Sky-Line of Central London. *Area*, 39.2: 195–205.

Cohen, Erik. 1979. 'A Phenomenology of Tourist Experience'. *Sociology*, 13(2), 179–201.

Collins, Timothy. 2008. 'Unevenness in Urban Governance: Stadium Building and Downtown Redevelopment in Phoenix, Arizona'. *Environment and Planning C: Government and Policy*, 26(6):1177–1196.

Daniels, Margaret. 2007. 'Central Place Theory and Sport Tourism Impacts'. *Annals of Tourism Research* 34(2), 332–347.

Davies, Larissa. 2005. 'Not in my Back Yard! Sports Stadia Location and the Property Market'. *Area*, 37(3), 268–276.

DCMS. 2012. *900K Overseas Visitors Watch Football*. Department for Digital, Culture, Media and Sport, last modified 22 October 2012. Available at https://www.gov.uk/government/news/900k-overseas-visitors-watch-football (accessed 16 January 2018).

Eurosport. 2015. The Global Game: The Premier League's International Reach, broken down'. *Eurosport*, last modified 11 August 2015, Available at https://www.eurosport.com/football/premier-league/2015-2016/the-global-game_sto4853526/story.shtml (accessed 19 February).

Flick, Uwe. 2009. *An Introduction to Qualitative Research*. London: Sage

Gammon, Sean 2004. 'Secular Pilgrimage and Sport Tourism'. In Brent Ritchie and Daryl Adair eds, *Sport Tourism, Interrelationships, Impacts and Issues*. Clevedon: Channel View Publications.

Gammon, Sean. 2011. 'Sporting's New Attractions? The Commodification of the Sleeping Stadium'. In Richard Sharpley and Philip Stone, eds, *Tourism Experience: Contemporary Perspectives*. Abingdon: Routledge.

Gammon, Sean and Victoria Fear. 2005. 'Stadia Tours and the Power of Backstage'. *Journal of Sport and Tourism*, 10(4), 243–252.

Gammon, Sean and Tom Robinson. 1997. 'Sport and Tourism: A Conceptual Framework'. *Journal of Sport and Tourism* 4(3), 11–18.

Gandomi, Amir and Murtaza Haider. 2015. 'Beyond the Hype: Big Data Concepts, Methods, and Analytics'. *International Journal of Information Management*, 35(2),137–144.

Gebauer, Gunter. 2010. 'Heroes, Myths and Magic Moments'. In Sybille Frank and Silke Steet, eds, *Stadium Worlds: Football, Space and the Built Environment*. Abingdon: Routledge.

Gibson, Heather. 1998. 'Active Sport Tourism: Who Participates?'. *Leisure Studies* 17(2), 155–170.

Gibson, Heather, Willming, Cynthia, and Andrew Holdnak. 2003. 'Small-scale Event Sport Tourism: Fans as Tourists'. *Tourism Management*, 24(2), 181–190.

Ginesta, Xavier. 2017. 'The Business of Stadia: Maximizing the use of Spanish Football Venues'. *Tourism and Hospitality Research* 17(4), 411–423.

Goffman, Erving. 1959. *The Presentation of Self in Everyday Life*. Garden City, NY: Doubleday.

Gratton, Chris and Ian Henry, eds. 2001. *Sport in the City: The Role of Sport in Economic and Social Regeneration*. London: Routledge.

Gray, Jonathan. 2003. 'New Audiences, New Textualities: Anti-Fans and Non-Fans'. *International Journal of Cultural Studies* 6(1), 64–81.

Greater London Authority. 2014. *London: Home of World-Class Sport*. London: GLA.

Gretzel, Ulrike and Kyung Hyan Yoo. 2008. 'Use and Impact of Online Travel Reviews'. In Peter O'Connor, Wolfram Höpken and Ulrike Gretzel, eds, *Information and Communication Technologies in Tourism 2008: Proceedings of the International Conference in Innsbruck, Austria, 2008*. Vienna: Springer Vienna.

Guest, Greg, MacQueen, Kathleen, and Emily Namey. 2011. *Applied Thematic Analysis*. London: Sage.

Hammond, Michael and Jerry Wellington. 2013. *Research Methods: The Key Concepts*. Abingdon: Routledge.

Herring, Rachel. 2004. 'Governance, Sport and the City: Using Case Studies to Inform Policy'. *Cross-National Research Papers* 7(5),17–26.

Higham, James. 2005. 'Sport Tourism as an Attraction for Managing Seasonality'. *Sport in Society* 8(2), 238–262.

Hinch, Tom and James Higham. 2005. 'Sport, Tourism and Authenticity'. *European Sport Management Quarterly* 5(3), 243–256. Hinch, Tom and James Higham. 2011. *Sport Tourism Development*. 2nd Edition, Bristol: Channel View Publications.

Horak, Roman. 1995. 'Moving to Suburbia: Stadium Relocation and Modernization in Vienna.' In John Bale, ed, *The Stadium and the City*. Keele: Keele University Press.

Hudson, Simon et al. 2001. 'Distribution Channels in the Travel Industry: Using Mystery Shoppers to Understand the Influence of Travel Agency Recommendations'. *Journal of Travel Research* 40(2), 148–154.

Hunt, Kenneth, Bristol, Terry, and Edward Bashaw. 1999. 'A Conceptual Approach to Classifying Sports Fans'. *Journal of Services Marketing* 13(6), 439–452.

John, Geraint, Sheard, Rod, and Ben Vickery. 2013. *Stadia: The Populous Design and Development Guide*. Abingdon: Routledge.

Jones, Calvin. 2001. 'A Level Playing Field? Sports Stadium Infrastructure and Urban Development in the United Kingdom'. *Environment and Planning A: Economy and Space,* 33(5): 845–861.

Jones, Ian. 2000. 'A Model of Serious Leisure Identification: The Case of Football Fandom'. *Leisure Studies,* 19(4), 283–298.

Jorgensen, Danny. 1989. *Participant Observation: A Methodology for Human Studies, Applied Social Research Methods*. London: Sage.

Jorgenson, Per. 1995. 'Copenhagen's Parken: A Sacred Place?'. In John Bale, *The Stadium and the City*. Keele: Keele University Press.

Joseph, Janelle. 2011. 'A Diaspora Approach to Sport Tourism'. *Journal of Sport and Social Issues* 35(2), 146–167.

Kellett, Pamm. 2007. 'Sport Museums: Marketing to Engage Consumers in Sport Heritage'. In Ruth Rentschler and Anne-Marie Hede, eds, *Museum Marketing: Competing in the Global Marketplace*. Oxford: Butterworth–Heinemann.

Kyte, Simon. 2012. *Tourism in London. Working Paper 53*. London: Greater London Authority.

Latuf de Oliveira Sanchez, Renata and Stephen Essex. 2017. 'The Challenge of Urban Design in Securing Post-event Legacies of Olympic Parks'. *Journal of Urban Design*, 23, 278–297.

LLDC. 2017. *Annual Report and Accounts 2016–2017*. London: LLDC.

London and Partners. 2016. *The Impact of Event Tourism on London's Economy*. London: London and Partners.

London and Partners. 2018. 'Major Sporting and Cultural Events' London and Partners. Available at http://www.londonandpartners.com/what-we-do/major-sporting-and-cultural-events (accessed 18 January 2018).

London Assembly. 2015. *The Regeneration Game: Stadium-led Regeneration*. London: London Assembly Regeneration Committee.

MacCannell, Dean. 1973. 'Staged Authenticity – Arrangements of Social Space in Tourist Settings'. *American Journal of Sociology* 79(3), 589–603.

MacCannell, Dean. 1976. *The Tourist: A New Theory of the Leisure Class*. New York, NY: Schocken Books.

Maguire, Joseph and Catherine Possamai. 2005. '"Back to the Valley": Local Responses to the Changing Culture of Football'. In Joseph Maguire, ed, *Power and Global Sport: Zones of Prestige, Emulation and Resistance*. London: Routledge.

Majendie, Matt. 2018, 'Crystal Palace athletics stadium faces last chance', *Evening Standard*, Last modified 13 September. Available at https://www.standard.co.uk/sport/athletics/crystal-palace-athletics-stadium-faces-last-chance-a3935096.html (accessed 5 November 2018).

Maitland, Robert. 2017. 'Cool Suburbs: A Strategy for Sustainable Tourism?'. In Susan Slocus and Carol Kline, eds, *Linking Urban and Rural Tourism: Strategies in Sustainability*, Wallingford: CABI.

Nelson, Arthur. 2001. 'Prosperity or Blight? A Question of Major League Stadia Locations'. *Economic Development Quarterly* 15(3), 255–265.

Ng Kwet Shing, Michelle and Laura Spence. 2002. 'Investigating the Limits of Competitive Intelligence Gathering: Is Mystery Shopping Ethical?'. *Business Ethics: A European Review* 11(4), 343–353.

Oliver, John and Keith Eales. 2008. 'Research Ethics: Re-evaluating the Consequentialist Perspective of using Covert Participant Observation in Management Research'. *Qualitative Market Research: An International Journal* 11(3), 344–357.

Paskaleva-Shapira, Krassimira. 2007. 'New Paradigms in City Tourism Management: Redefining Destination Promotion'. *Journal of Travel Research,* 46(1), 108–114.

Pearce, Philip and Gianna Moscardo. 1986. 'The Concept of Authenticity in Tourist Experiences'. *The Australian and New Zealand Journal of Sociology,* 22(1), 121–132.

Pedro, Dionísio, Carmo, Leal, and Moutinho Luiz. 2008. 'Fandom Affiliation and Tribal Behaviour: A Sports Marketing Application'. *Qualitative Market Research: An International Journal* 11(1), 17–39.

Price, Joseph. 2001. 'Fervent Faith: Sports as Religion in America'. In Joseph Price, ed, *From Season to Season: Sports as American Religion.* Macon, GA: Mercer University Press.

Punch, Keith. 2005. *Introduction to Social Research: Quantitative and Qualitative Approaches.* 2nd edition. London: Sage.

Ramshaw, Gregory and Sean Gammon. 2010. 'On Home Ground? Twickenham Stadium Tours and the Construction of Sport Heritage'. *Journal of Heritage Tourism,* 5(2), 87–102.

Ramshaw, Gregory, Gammon, Sean, and Wei-Jue Huang. 2013. 'Acquired Pasts and the Commodification of Borrowed Heritage: The Case of the Bank of America Stadium tour'. *Journal of Sport and Tourism,* 18(1), 17–31.

Redmond, Gerald. 1973. 'A Plethora of Shrines: Sport in the Museum and Hall of Fame.' *Quest* 19(1), 41–48.

Rugby Football Foundation. 2015. *Report and Financial Statements.* Available at http://www.englandrugbyfiles.com/rff/images/reports/RFF%20Financial%20 Statements%202015.pdf (accessed 7 November 2018).

Russo, Antonio Paolo. 2002. 'The "Vicious Circle" of Tourism Development in Heritage Cities'. *Annals of Tourism Research,* 29(1), 165–182.

SCC. 2016. *Surrey County Cricket Club: Annual Report and Accounts.* Available at https://www.kiaoval.com/wp-content/uploads/2013/05/SCCC-Annual-Report_2016-PRINT-READY.pdf (accessed 7 November 2018).

Sharpley, Richard. 2008. *Tourism, Tourists and Society,* Fourth Edition. Huntingdon: ELM Publications.

Slack, Trevor and John Amis. 2004. '"Money for Nothing and Your Cheques for Free?" A Critical Perspective on Sport Sponsorship.' In Trevor Slack, ed, *The Commercialisation of Sport.* London: Routledge.

Smith, Andrew. 2005. 'Conceptualizing City Image Change: The 'Re-Imaging' of Barcelona.' *Tourism Geographies* 7(4), 398–423.

Sport Business. 2016. 'Media Rights Value in the Top European Football Leagues, 2016–17.' SBG Companies, last modified 23 September 2016. Available at https://www.sportbusiness.com/tv-sports-markets/media-rights-value-top-european-football-leagues-2016–17 (accessed 19 February 2018).

Sportcal. 2017. *Global Sports Impact Report.* London: Sportcal Global Communications.

Stevens, Terry and Glyn Wootton. 1997. 'Sports Stadia and Arena: Realising Their Full Potential'. *Tourism Recreation Research* 22(2):49–56.

Sutton, William et al. 1997. 'Creating and Fostering Fan Identification in Professional Sports'. *Sport Marketing Quarterly* 6(1):15–22.

Tallentire, Philip. 2018. 'Moving Home's Traumatic: Why Middlesbrough got it so right and West Ham so wrong'. *Gazette Live*, 15 March, Sport. Available at https://www.gazettelive.co.uk/sport/football/football-news/moving-homes-traumatic-middlesbrough-right-14396216 (accessed 16 March 2018).

Thornley, Andy. 2002. 'Urban Regeneration and Sports Stadia'. *European Planning Studies* 10(7), 813–818.

Vamplew, Wray et al. 1998. 'Sweet FA: Fans' Rights and Club Relocations'. *Occasional Papers in Football Studies* 1(2), 55–68.

Visit Britain. 2015. *Football Tourism scores for Britain*. London: Visit Britain.

Visit Britain. 2017. 'Latest Quarterly Data by Area.' Available at https://www.visitbritain.org/latest-quarterly-data-area (accessed 18 January).

Visit England. 2016. 'Annual Survey of Visits to Visitor Attractions'. Available at https://www.visitbritain.org/annual-survey-visits-visitor-attractions-latest-results (accessed 16 January).

Wann, Daniel and Nyla Branscombe. 1993. 'Sports Fans: Measuring Degree of Identification with their Team'. *International Journal of Sport Psychology* 24(1), 1–17.

Weed, Mike and Chris Bull. 2004. *Sports Tourism: Participants, Policy and Providers*. Oxford: Butterworth Heinemann.

Wigoder, Meir. 2002. 'The 'Solar Eye' of Vision: Emergence of the Skyscraper-Viewer in the Discourse on Heights in New York City, 1890-1920.' *Journal of the Society of Architectural Historians* 61(2): 152–169.

Williams, Colin. 1997. *Consumer Services and Economic Development*. London: Routledge.

Williams, John. 2007. 'Rethinking Sports Fandom: The Case of European Soccer'. *Leisure Studies* 26(2), 127–146.

Wilson, Alan. 1998. 'The Role of Mystery Shopping in the Measurement of Service Performance'. *Managing Service Quality: An International Journal* 8(6), 414–420.

Wright, Richard Keith. 2012. 'Stadia, Identity and Belonging'. In Richard Shipway and Alan Fyall, eds, *International Sports Events: Impacts, Experiences and Identities*. Abingdon: Routledge.

Zeng, Guojun, Go, Frank, and Christian Kolmer. 2011. 'The Impact of International TV Media Coverage of the Beijing Olympics 2008 on China's Media Image Formation: A Media Content Analysis Perspective'. *International Journal of Sports Marketing and Sponsorship* 12(4), 39–56.

Zinganel, Michael. 2010. 'The Stadium as Cash Machine'. In Sybille Frank and Silke Steets, eds, *Stadium Worlds: Football, Space and the Built Environment*. Abingdon: Routledge.

Vertical City Tourism: Heightened Aesthetic and Kinaesthetic Experiences

Andrew Smith

Introduction

Urban areas have traditionally been analysed as two-dimensional phenomena, with emphasis placed on the spatial distribution of features, connectivity at ground level and horizontal urban expansion. This neglects the verticality of cities – arguably their defining feature – which has become even more significant as more and more tall buildings are constructed (Graham and Hewitt 2012). In 2000 there were 265 buildings in the world that were over 200 metres tall (CTBUH 2016). By 2010 this had risen to 612 and the latest figures suggest there are now 1,169 buildings that exceed this height – a 441 per cent increase since the Millennium (CTBUH 2016). This growth has been accompanied by calls for more recognition of the verticality of urban space (Graham and Hewitt 2012; McNeill 2005), and, in recent years, academics from various disciplines have responded to these calls (Deriu 2018). Much of this emerging body of work is linked to urban militarisation, securitisation and surveillance but, as Harris (2015, 604) notes, it is important to recognise other types of 'vertical forms, landscapes and experiences', including those that involve the 'production, marketing and commodification of urban views'.

How to cite this book chapter:

Smith, A. 2019. Vertical City Tourism: Heightened Aesthetic and Kinaesthetic Experiences. In: Smith, A. and Graham, A. (eds.) *Destination London: The Expansion of the Visitor Economy.* Pp. 117–139. London: University of Westminster Press. DOI: https://doi.org/10.16997/book35.f. License: CC-BY-NC-ND

Published research on city tourism also tends to neglect the verticality of urban destinations. Even in the rare instances where the tourism implications of tall buildings have been analysed, the focus has been on their role as traditional attractions or their contribution to the general urban milieu (Leiper and Park 2010). By focusing on these aspects, accounts tend to be overly negative with Leiper and Park (2010, 347) arguing that 'skyscrapers are not merely deficient as attractions, they reduce the attractiveness of cities for many tourists'. This restricted perspective ignores the sights and feelings tourists can experience by 'getting high'. Tall urban structures do not merely provide things to be seen: they are 'machines for seeing' (Wigoder 2002), and they provide opportunities to descend, ascend and traverse the built environment. These neglected aspects of vertical urban tourism are discussed here.

The aim of this chapter is to examine the increasing amount of opportunities to experience London from up high. This trend is driven by the increased number of high rise towers that have been built in London since the Millennium (Charney 2007; Clark 2015), but also by the growth of purpose-built attractions which trade on the value of London's cityscape (e.g. The London Eye and Ancelor Mittal Orbit). Tourists visiting London have always been attracted to high points from which the city can be viewed, but this chapter analyses new opportunities and more diverse ways that London's tourists can 'hit the heights'. The chapter also explores how passive forms of consumption have been supplemented by new attractions that facilitate active engagement. In the twenty-first century tourists are not merely able to access great views, they are now able to experience height in a more embodied sense by climbing up, riding on, and sliding down, tall structures. This trend is linked to increased demand for more active forms of tourism; and the ways that adventure tourism – normally something associated with rural contexts – is increasingly being offered in city centres (Beedie 2005). New high rise attractions extend the types of experiences offered by London and expand the city's tourism territory vertically.

This chapter is based on, and builds on, the work of Davide Deriu, a Reader in Architectural History and Theory at the University of Westminster, who has written extensively about the significance of aerial views and the development of new experiential forms of high-rise architecture. Davide led the innovative, multidisciplinary Vertigo research project which provided the inspiration for this chapter and his conceptualisation of the shift from 'architectures of vision' to 'architectures of experience' (Deriu, 2018) is adopted as one of the key ideas used to frame the text.

A Short History of Vertical Urban Tourism

In his comprehensive review of vertical urbanism(s) Harris (2015) highlights four models which reflect different periods of high rise construction. Early projects were based on spiritual ambitions – elevated churches and temples were constructed, not merely to assert the power of religious institutions, but to connect urban populations to the heavenly sky. The late nineteenth century spawned a second model involving the construction of monuments to corporate capitalism – i.e. skyscrapers. This was followed in the mid-twentieth century by modernist landscapes of high rise housing introduced by Le Corbusier and other proponents of building 'streets in the sky'. More recently we have seen the rise of a new breed of high rise urbanism in global Asia, a trend which has shifted the focus of verticality eastwards. Each of these models has contributed to the verticality of contemporary urbanism, particularly in global cities like London. Indeed, Harris (2015) notes that all four models are evident in London's contemporary cityscape, citing St Paul's Cathedral, Adelaide House, The Barbican, and the plethora of new towers funded by Asian investors, as examples.

Recognising the historical evolution of vertical urbanism allows us to better understand the allure of the city panorama; a view that is extensive, unbroken and multi-directional. People have long wanted to consume cities as holistic landscapes via high points located either inside or outside the city boundaries. The contemporary popularity of these urban panoramas is usually traced back to renaissance cities and the production of town views from elevated vantage points (Balm and Holcomb 2003). As Hinchcliffe and Deriu (2010, 221) remind us, 'once you could climb the cathedral towers of any European city and view its whole extent'. This tradition inspired a whole range of artistic outputs – including paintings, photographs, poems and novels – reimagining the city as 'a work of art' (Olsen 1988; Boyer 1994), and whetting the appetite for elevated views in the contemporary era.

Various new technologies facilitated the rise of the panoramic views in modern cites (i.e. post 1851). In her wonderful account of the growth of tourism in US cities, Cocks (2001) highlights how emerging forms of urban transportation allowed cities to be consumed panoramically. Initially, this experience was facilitated by the streetcar or trolley bus, vehicles which offered an elevated view of the city. These experiences were assisted by guides and guidebooks, media that 'directed the attention of the car's riders to the historical and aesthetic features of the landscape' (Cocks 2001, 167). Tourists were thus taught what to see, and how to see it, an education which helped to reinforce the significance of the urban panorama. In what would become a familiar explanation for the popularity of viewing platforms, transport vehicles offered visitors the chance to consume the city as a spectacle, rather than as a direct experience. The elevated position not only provided a better view, it differentiated tourists from citizens – ensuring leisure visitors were not mistaken for the leisured poor (Cocks 2001).

The other new technologies that allowed tourists to consume cities from above were the elevator, the steel frame, and the related rise of the modern skyscraper. Although the first skyscraper and the first Ferris wheel were built in Chicago, high rise tourism first flourished in New York. In the period 1870–1910 New York's skyline was transformed into 'a spectacle of skyscrapers' – making this rapidly developing city 'one of the modern wonders of the touristic world' (Gilbert and Hancock 2006, 90). Spending time on the rooftops of New York became a popular pastime in this period, with high rise buildings not merely offering spectacular views, but a welcome chance to escape the pollution and the crowds of the streets below. For the first time, people didn't need to undertake a lengthy journey to exit the city, they could achieve a vertical escape simply by pressing an elevator button (Wigoder 2002).

People had to be trained to see the urban beauty of the modern city (Cocks 2001) and links to natural landscapes were made to convince people that urban panoramas were worth seeing. In the 18th and 19th Centuries, the rise of romanticism encouraged scenic tourism and the appreciation of impressive natural scenery (Urry 1990). This was translated into an urban context via the provision of spectacular views. According to Wigoder (2002, 159) new skyscrapers 'offered the possibility of standing at the edge of the roof and looking down at the city as if it were a sublime, romantic view enjoyed from a mountain crag'. Tourists were still uncertain about the aesthetic value of modern buildings, but viewed from above these merged together to form a spectacular cityscape. The way panoramas naturalise the city by turning it into a landscape is noted by Barthes (1983) in his famous account of the view from atop the Eiffel Tower. It is also reaffirmed in contemporary accounts which suggest that aerial perspectives transform streets into canyons (Deriu 2016).

The construction of the Empire State Building in the 1930s marked a new phase of high rise tourism. As Gilbert and Hancock (2006, 93) identify: 'unlike earlier skyscrapers that had become touriot attractions, the Empire State Building was consciously designed with tourism in mind'. Purpose built 'observatories' were constructed on the 86th and 102nd floors with dedicated lifts for the visitors who wanted to enjoy the view (MacCannell 1999). The Empire State Building tends to be cited as an iconic structure to look *at*; but this was a pioneering example of a place to look *from*. Two years after the Empire State Building's observatories opened in 1931, tourists could also enjoy the view from the newly constructed Rockefeller Centre observation deck on the 70th floor of the RCA Tower. Even though visitors had to pay to enter, 1,300 people a day were visiting by 1935 making it the top New York destination for 33 per cent of all visitors (London 2013). This space was designed to evoke the deck of an ocean liner (hence it was called an observation deck), connoting this was a luxury experience and one that transformed the city below into an undifferentiated sea. These pioneering examples have inspired a high range of observation decks and observatories throughout the world, including others built atop skyscrapers, but also observation towers featuring revolving restaurants.

The Allure of the Panoramic View

Various authors have explored the appeal of the city viewed as a panorama from on high. As Dorrian (2009) notes, to go up is to see more, but it is also to see in a different way. Many accounts use religious analogies to explain the appeal of this alternative perspective, with aerial views associated with transcendence, levitation, omnipotence and the scopic power of a god's-eye view

(Dorrian 2009). Humans seem to have an insatiable urge to encapsulate the city as a whole 'unit' (Wigoder 2002), or to read the city like a text (De Certeau 1984), and these interpretations also help to explain the enduring appeal of panoramic views. Through abstraction the city becomes more comprehensible (Jansson and Lagerkvist 2009), something that provides reassurance and comfort. An elevated vantage point allows people to appropriate the city as an object and this is further enabled by photographing the view – an activity that dominates the contemporary experience of panoramic viewpoints. A slightly contradictory interpretation is that people are awed by the spectacle of infinity and immensity that aerial views provide (Dorrian 2009). This suggests urban panoramas can also be understood via reference to Kant's interpretation of the sublime – the experience of something beyond conceptualisation which makes us realise our physical impotence.

Being high up in the city is associated with authority, status and exclusivity and these connotations also help to explain the allure, but also the wider implications, of views from above. The skyscraper is regarded by some as a metaphor for the stratification of the contemporary city, with the most affluent living at the top and the poor living at the bottom (the underclass presumably resides in the basement). Just as citizens seek upward mobility, tourists welcome the chance to rise above the chaos and poverty of the city and experience it from on high. Tourists are attracted to cities but they also want to escape from them. They want the best of both worlds – to exist simultaneously within and outside cities – and high rise buildings (and urban parks) provide such opportunities. This interpretation is particularly relevant to tourists visiting developing world cities, where verticality is coveted as it provides security from the perceived insecurities below. For example, Wharton's (2001) history of the Hilton Group shows how this company's high rise hotels allowed tourists to consume foreign territories from safe sites.

If 'getting high' is a vehicle through which to achieve control, abstraction and exclusivity, then it is about power. This is a key theme in much of the literature on city panoramas and it is particularly relevant to the tourism-focused discussion here. Thanks to John Urry's acclaimed work, the tourist gaze is understood as an expression of power. By consuming and prioritising signs, tourists exert influence over the people, cultures, sites and objects that are gazed upon. The powerful objectification of the tourist gaze is a function of the distance and detachment of the tourist from the objects they are consuming – and by ascending tall structures the tourist is able to achieve distance and separation. Therefore, the view from high above the city provides a particularly potent form of the tourist gaze.

Whilst it is important to acknowledge the interpretation of the panoramic view as an expression of power, it is also worth noting the counter arguments to this established position. Jansson and Lagerkvist (2009) challenge the idea that panoramas are inherently vehicles for promoting encapsulation and detachment. These authors argue that attempts to encapsulate cities need to be considered alongside the inevitability of decapsulation – where the magic of the spectacle is broken and replaced with fear and boredom. This interpretation reflects other critical accounts which also challenge the idea that people gain reassuring control over cities via aerial views. Dorrian (2009) suggests that being above things can be disconcerting, because of the way the ground appears to dissolve and because urban features seem to merge into each other. In such instances people may suffer the despair of not knowing what is significant and what is not (Dorrian 2009). Contemplating the immensity of the contemporary city can also involve a crushing and decentring diminishment of ourselves (Dorrian 2009). These negative aspects of consuming cities do not necessarily reduce their appeal as attractions: the enduring popularity of theme parks, adventure tourism and dark attractions highlight that some tourists are attracted to disorienting, scary and disturbing experiences.

The unsettling effect of viewing a city from above can be better understood by exploring the notion of vertigo. This is a physical and psychological condition, but the term is now also deployed metaphorically to refer to the nervous instability people feel in the modern city (James 2013; Deriu 2018). Vertigo is used colloquially to refer to the unease felt when looking down from great heights, but as a medical condition it is defined as dizziness – a sensation of giddiness and disorientation caused by problems with balance mechanisms in the inner ear. The derivation of the word comes from the Latin *vertere* – to turn – and there are etymological links to the words whirl, whirlpool and vortex. Recognising the physical condition of vertigo is important in the context of this chapter, as it reminds us that experiences atop high rise structures stimulate physical sensations, rather than merely visual ones (Deriu 2018). This helps us to understand the recent changes made to traditional observation decks and viewing platforms – such as adding slides and transparent floors (Deriu 2018). As Deriu (2018) notes, these can be understood as attempts to develop the physical dimension of these attractions, shifting the focus from aesthetics to kinaesthetics.

Seeing London Differently

The extended introduction above provides the historical and conceptual context for this chapter. Subsequent sections focus on opportunities to consume London from above: by examining the development of new viewing platforms; and then by exploring the way these attractions have been supplemented by more dynamic experiences. These allow tourists to enjoy panoramic views whilst ascending, descending or traversing high rise structures.

New Opportunities to Consume London Passively from Above

London has always attracted tourists wishing to view the city from above. The physical geography of the city allows views of central areas: for example, from

Forest Hill and Greenwich Park in the south, and from Parliament Hill and Alexandra Palace in the north. Tourists and residents can still enjoy these views today – something which has been achieved through innovative planning controls introduced in 1991. London now has a list of 'protected vistas' which prevents new development blocking visual corridors – mostly views from peripheral parks to St Paul's Cathedral and/or the Houses of Parliament. This means visitors can view panoramas of London from its elevated suburbs, as well as from tall buildings in the city centre. However, these protected vistas are currently being challenged by the large volume of high rise building planned for and already built in central London. For over 250 years (1710–1964), the city's tallest building was St Paul's Cathedral, and visitors have long climbed the stairs to view the city from the roof. However, the construction of the BT Tower, CentrePoint and the NatWest Tower in the 1960s and 1970s started a trend of verticalisation, and this has intensified in recent years. Since 2000, multiple tall buildings have been constructed, particularly in East London (at Canary Wharf) and in the City of London itself. Care has been taken to ensure historic buildings are not crowded out by these new towers, but London's character as a relatively low rise city compared to other World Cities is beginning to disappear. This trend is set to continue: London's housing shortage has inspired a new phase of vertical development and, at the time of writing, 455 new buildings of over 20 storeys are planned (NLA 2017).

Alongside housing London's growing population, the main justification cited for developing new tall buildings in London is the need to provide new office space to ensure London remains one of the world's most significant centres for financial services (Clark 2015). In the period 2000–2008 a powerful coalition involving The Mayor of London, the National Government (more specifically the Office of the Deputy Prime Minister) and various property interests used this rationale to push through a number of controversial projects. Companies were threatening to leave London unless they were permitted to build new high spec office space (Charney 2007) and London's first elected Mayor

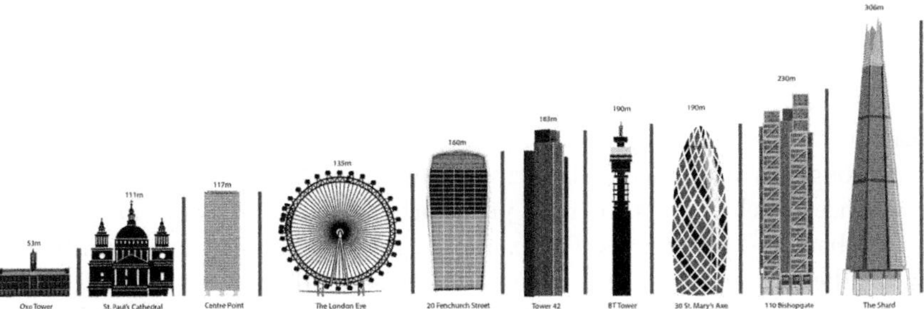

Figure 6.1: Hitting the Heights – A Height Chart illustrating London's Notable Tall Buildings (© Mason Edwards).

(Ken Livingstone) was eager to ensure this did not happen. Resistance was diluted by using high quality designs that were both eye-catching and more environmentally sustainable (Charney 2007). The popular and critical acclaim attained by early examples – such as Norman Foster's 'Gherkin' (30 St Mary's Axe) – provided a convenient justification to build more towers. Many of these later projects have been less well received: for example, the 'Walkie Talkie' (20 Fenchurch Street) designed by Rafael Vinoly is an ugly and imposing structure which has upper floors that are more voluminous than the lower ones. The zenith of London's post–2000 shift upwards was the construction of The Shard just outside the City of London (32 London Bridge Street). This structure was designed by Renzo Piano as 'a vertical city' and it hosts residential apartments, office space and a hotel. Standing over 1,000 feet tall, this is the tallest building in Europe.

London's new generation of high rise towers have provided opportunities for new visitor experiences. In several instances (e.g. The Shard, and 20 Fenchurch Street), viewing platforms were included in the designs that allow the public to experience open air views from the upper floors. In the case of The Shard, tickets are expensive, with adults currently charged over £30 to access an attraction branded 'The View from The Shard' on the 69th and 72nd Floors. Various events are staged to encourage additional demand and repeat visits, including several that capitalise on the spiritual and romantic connotations of panoramic views. For example, an all-night music event is staged on the eve of the summer solstice which allows revellers to watch the sunrise over the city. The View from the Shard has also become a place associated with love and romance; on Valentine's Day 2015 over six thousand people visited, the highest amount ever recorded on a single day (The Shard 2016).

The viewing area which opened in January 2015 at 20 Fenchurch Street is a different type of attraction than The View from the Shard. Here, developers were required to provide an accessible public space at the top of their building in order to gain planning permission. This means access is free, although visitors have to book in advance and endure arduous security checks to enter. These are not the only criticisms of the project: the space is promoted as 'The SkyGarden' and is meant to be a public garden, but it feels more like a hotel lobby than a public space. The limited dimensions mean it has been dubbed The SkyRockery by critics (Wainwright 2015). Nevertheless, The SkyGarden has proved to be extremely popular: 1,210,049 people visited in the first two years it was open (2015–2017) (Gillespies 2017).

Although formal viewing platforms are not provided in London's other high rise office developments, there is public access of sorts via the provision of hotels, bars and restaurants. For example, the general public are permitted to access the restaurants on the 38th and 40th floors of the new tower built at 110 Bishopsgate (formerly the Heron Tower, now rather depressingly called the Salesforce Tower). Fine dining with panoramic views has become an important part of the skyscraper experience, and is something that reflects and reinforces

Figure 6.2: The View from the SkyGarden – 20 Fenchurch Street (Photo: Tristan Luker).

the reputation of high rise buildings as exclusive territories. A viewing platform is clearly tourist terrain, but by eating in a restaurant or drinking champagne in a bar visitors get a chance to mix with city professionals and sample the 'high life' enjoyed by urban elites.

The penthouse epitomises exclusive urban living, and tourists can also experience what it would be like to wake up with panoramic views of London by staying in one of London's new high rise hotels. For over 50 years The Park Lane Hilton, was London's tallest hotel (101 metres), but this accolade is now held by the Novotel Canary Wharf which opened in in 2017 (127 metres). Other mixed use towers also offer hotel accommodation. Affluent visitors can experience 'a new level of luxury' by staying in the Shangri-La Hotel – located on floors 34 to 52 of The Shard – where prices for 'rooms with a view' start at £496 a night. One review of the Shangri-La which features prominently on the Hotel's website reaffirms why people want stay in such accommodation: 'From up here the frantic city seems so serene' (Financial Times 2014). Alongside the panoramic views, this desire to escape the street and access elevated sanctuary helps us to explain the appeal of high rise structures.

The appearance of new viewing platforms in London has been driven by the construction of new high rise office towers, but it also results from the regeneration of historic structures and the provision of new tourist attractions. Amongst the most popular elevated viewing points along the River Thames are Tate Modern and The Oxo Tower. These buildings were originally built in the 1930s/1940s as industrial installations and both were regenerated in the 1990s as part of the transformation of the South Bank of the Thames. They are now open to the public and provide opportunities to view London from the upper floors. Whilst 8–10 storey structures may not provide the spectacular panoramas offered by skyscrapers, they offer elevated views where observers can engage with people below (Deriu 2018). Tate Modern has recently been extended by adding a 200-ft high pyramid at the back of the original building. Switch House offers views of the city via a roof terrace, but its proximity to new high rise residential development next door has caused some unexpected problems. Instead of admiring the views across the river and the rooftops, visitors have been staring into the new glass-walled apartments opposite. This adds a whole new dimension to the argument that elevation turns viewers into distanced voyeurs (Wigoder 2002). Conflict between different users of high structures also reminds us that we need to understand the relationship *between* high rise buildings rather than analysing them as stand-alone structures.

Perhaps the most famous way of seeing London from above is by riding the London Eye, the enormous Ferris wheel installed on the South Bank of the Thames close to Westminster Bridge. The London Eye opened in 2000 and was initially sanctioned as a temporary attraction but its success meant it was retained as a permanent structure. The Eye's popularity has endured and it remains the most popular paid-for visitor attraction in the UK, encouraging other cities to construct similar structures. The attraction was originally sponsored by British Airways and experiences were promoted as 'flights', emphasising the dynamic aerial views offered. Ferris wheels provide a different type of high rise experience as they provide panoramic views that change as passengers are transported around the circumference of the wheel. The design of the cabins means that views are framed into pictorial compositions, turning the city panorama into a series of artworks (Borden 2014). Borden (2014) also suggests that Ferris wheels act as time machines, not just because of their clock-like circular movement, but because of their historic significance. The view from above is often regarded as an opportunity to glimpse into the future, but the appeal of Ferris wheels is very different – they stimulate feelings of nostalgia and reconnect us to the technologies of the past (Borden 2014).

Like other high rise structures, The London Eye consciously separates people from the surrounding city. Transportation via sealed capsules disconnects the observer from their external environment, with the sounds of the city silenced and the possibility of encountering strangers removed. For 30 minutes tourists are able to enjoy views of the city without having to encounter the city itself. Passengers on the London Eye are thus 'lifted out of the city's grasp' (De Certeau 1984, 92), allowing them to simultaneously escape the city whilst giving them more power and control over it. As Barthes (1983, 250) notes, when you ascend a structure like the London Eye or the Eiffel Tower 'one can feel cut off from the world and yet the owner of a world'. By abstracting the city into a map,

Figure 6.3: The London Eye (Photo: Tristan Luker).

miniature and model, the London Eye experience allows passengers to own London (Borden 2014; Dorrian 2009).

In the early days of modern tourism, people visiting London were keen to view the city panoramically from vehicles that offered physical separation and elevation. For example Barton (1996) notes that omnibuses were the transport of choice for Indian travellers to London in the nineteenth century because they offered a bird's-eye view. Open top, double decker buses remain a popular way of consuming London today – like other viewpoints they offer the elevation and protection sought by less adventurous tourists. Panoramic views from vehicles are also provided by new additions to London's transport infrastructure. In 2012, a new cable car over the River Thames opened connecting North Greenwich and Canning Town. Cable cars are normally associated with mountainous landscapes, so their introduction to London represents a further example of the way rural attractions and adventure tourism are increasingly urbanised. This new way of crossing the river is an integrated part of London's transport network but it is sponsored by Emirates and promoted as an 'airline' – emphasising the way it offers tourists elevated views of the city. Despite the generous sponsorship deal, it is still subsidised by Transport for London and there are ongoing concerns about its long term viability. The peripheral location means it struggles to attract many commuters or tourists and, despite the very reasonable prices, the cable car is only used by approximately 1.5 million people every year (Transport for London 2017). However, the Emirates Airline improves accessibility to and from one of London's most deprived Boroughs (Newham) which suggests subsidies might be justified. Examples from further afield (e.g. Medellin) show that cable car technologies can improve mobility opportunities for some of the poorest citizens – a useful reminder that the vertical expansion of the city doesn't have to favour the rich and powerful (Brand and Davila 2011).

To comprehend the contemporary urban landscape, ascending tall structures is not enough – we need to fly (Hinchcliffe and Deriu 2010), and alongside simulated flight experiences – e.g. the London Eye or the Emirates Airline – real flight across London is an increasingly common way of experiencing the city from above. Millions of international tourists every year experience London vistas when they fly into one of the city's airports (especially London City Airport) see Chapter 5; and a more intimate version of this experience is now offered through helicopter tours. Prices in London start at £150 for flights lasting a mere 18 minutes, so this is an expensive experience and one that reaffirms the established link between urban elevation and exclusivity. These tours are linked to the rise of new residential towers in London as a new high rise tower in Battersea provides a convenient place to take off and land. Again, this highlights the relationships that exist between the different aspects of verticality emerging in contemporary London. In the future there are likely to be more opportunities to move between tall buildings without engaging with the street. The idea of urban elites travelling between high rise residences, hotels

Figure 6.4: The Emirates Air Line – London's New Aerial River Crossing (Photo: Eman Mustafa).

and offices via helicopters and never touching the ground seems like a dystopic vision from a J.G. Ballard novel, but it is already a reality in some South American cities (Graham and Hewitt 2012; Harris 2015).

New Ways of Consuming London Actively from Above

The previous section demonstrates the range of new opportunities to view London from above that have accompanied the city's recent verticalisation. Elevated positions provide great views, but attractions in London have also begun to offer more adventurous experiences which capitalise on the thrills of ascending, descending and traversing high places. As McKay (2013) identifies, tourists are no longer content with sightseeing or exploring passively; they want to experience urban areas whilst engaged in adrenalin rush activities. Several authors (Swarbrooke et al. 2003; Beedie 2005) also note that urban areas provide a new frontier for adventure tourism – something traditionally associated with natural landscapes. Adventure tourism is moving into cities and city tourism is moving into adventure, and the result is more adventurous urban destinations. Over the past few decades various adventure tourism activities, e.g. climbing and skiing, have been commodified and urbanised by the introduction of indoor facilities (Beedie 2005). But, more recently there has been an expansion in the number of adventure sports offered outside in less contrived settings, where the city is reimagined as an active landscape. A pioneering example was the bungee jump performed by members of Oxford University's Dangerous Sports Club from the Clifton Suspension Bridge in 1979. In the contemporary city, vertiginous adventure tourism is not merely confined to bungee jumping: climbing, abseiling, urbex (urban exploring) and free running are also examples of activities that make use of the vertical built environment. Participants are seeking various thrills, but are also looking to experience 'flow' – an ecstatic feeling linked to immersion in the moment where a person achieves a state of detachment from material reality (McKay 2013).

One way that city destinations have catered for the demand for more adventurous experiences is by adapting existing attractions, and Deriu (2018) examines the way viewing platforms have been updated to encourage more physical experiences. Following the example of Toronto's CN Tower, many observation decks have been fitted with transparent floors to add an element of excitement and danger. This type of attraction – which is a natural extension of the installation of glass lifts, glass staircases and other transparent ways of ascending built structures – has also been introduced in London. In 2014, the walkway that connects Tower Bridge's famous towers was fitted with a glass floor which allows people to look down at the vehicles and people crossing below. Most elevated viewpoints offer panoramas across the city, but Tower Bridge now offers a downwards view where urban features are seen directly from above. The introduction of a more experiential dimension means this heritage attraction is now promoted as the Tower Bridge *Experience*, refreshing its image and attracting a different audience. Nevertheless, this is relatively tame fare compared to other examples where the thrill of looking down from a tall building is exaggerated by structural transparency. For example, at the John Hancock Tower in Chicago visitors are invited to enter glass boxes that are tilted 30 degrees over a 300m drop. Deriu (2018) suggests these types of features exemplify the shift toward experience design in architecture, highlighting a shift from 'architectures of vision' to 'architectures of vertigo'.

Visitors usually ascend high rise attractions by taking lifts to upper floors. However, following the trend for more participatory and active experiences, there are now opportunities to climb London's vertical landscape. For example, an observation platform and a climbing route were recently installed on top of the O2 – one of London's most famous new buildings which now hosts the world's most popular indoor music venue. 'Up at the O2' opened in 2012 and it allows visitors to climb a tensile walkway to reach the top of this dome shaped structure 52 metres above ground. The notion of urban adventure tourism is explicitly acknowledged at this new attraction which is positioned as a 'mountaineering expedition' with visitors invited to start their journey at 'Base Camp' and then 'Conquer the Summit of London'. This attraction illustrates the trend for more physical experiences, and a desire for attractions that offer the excitement and spontaneity that is missing from quotidian life (Beedie 2005). The potential to 'climb an icon' at the O2 also highlights new demand for architecture that engages the public beyond the stimulation of their visual senses. Allowing people to climb buildings can create feelings of attachment and ownership – making architecture feel more public (Smith and Strand 2011). This is also part of the rationale for 'rooftopping' – where urban adventurers climb the vertical city not merely for the thrill of it, but in order to appropriate buildings, sabotaging 'the culture of passive consumption that underlies the society of the spectacle' (Deriu 2016, 1044).

Providing spectacular descents is an obvious way that high rise structures can cater for tourists seeking thrilling experiences. In 2007 the artist Carsten Holler caused a stir in London with his Tate Modern exhibition featuring a series of slides which transported people from upper levels to the floor of the Turbine Hall. This exhibition was called Test Site and Holler felt that his structures were prototypes for slides that could be introduced as permanent features of London's cityscape. Nine years later, this futuristic vision came a step closer when a slide he designed was installed on the Arcelor Mittal Orbit. The Orbit is a sculpture which was designed by the sculptor Anish Kapoor to provide London's Olympic Park with the iconic structure that Boris Johnson (Mayor of London 2008–2016) felt it lacked. An observation deck and lift had already been installed near the top to encourage people to ascend, but when The Orbit opened to the public visitor numbers were disappointing. Rather than closing the attraction, officials decided to reinvent it by adding an experiential

Figure 6.5: Up at the O2 – Greenwich Peninsula (Photo: Andrew Smith).

Figure 6.6: The Ancelor Mittal Orbit featuring a New Slide (Photo: Tristan Luker).

dimension and the structure that was installed means people can now descend the UK's tallest sculpture by travelling down the world's longest tunnel slide. Even though the slide is enclosed, several transparent sections mean that visitors can see London as they descend – producing an accelerated panoramic view. The overall effect is one of 'delightful terror', a defining characteristic of commodifed adventure where the hint of danger is combined with the knowledge that no harm will come (Beedie 2005).

The slide has stimulated new interest in visiting The Orbit. This attraction made a loss of £500,000 in 2015/6, but after the slide opened in June 2016 it returned a profit of over £100,000 during the rest of that year (The Wharf 2016). The Orbit's revised design and new-found popularity provides a clear demonstration of the need for twenty-first century viewing platforms to offer more than just views. Competition from other high rise structures and the appetite for more physical experiences is now forcing rival commercial viewing platforms to reinvent their attractions. In a direct response to the new threat posed by The Orbit, The View from The Shard is now augmented with virtual reality experiences that allow visitors to feel like they are sliding down from the top of the building or balancing along elevated steel frames.

Several authors, most notably Stevens (2007), have noted recent efforts to make our cities more playful. This does not just mean providing more opportunities for children, as playfulness is also something that is increasingly attractive to adults too. The introduction of The Orbit's slide is one example of this trend, but there are others too, with several other attractions trying to combine playfulness, adventure and panoramic views. For example, in 2017 a very long, very fast and very high zip wire was installed in Archbishop's Park in South London by Zip World, a company which normally operates in the Welsh countryside. This park was deliberately chosen to host the wire as it offered views of the Houses of Parliament, The London Eye and the River Thames. The essential appeal of the attraction is based on the way it combines speed, height and views:

> Get ready for the ride of your life on the fastest city zipwire! You will be taking off from a height of 35 meters (100 feet), that's more than 9 double decker buses! Catch never-before-seen views of the London's iconic skyline including Big Ben!
>
> (Zip World London 2017)

This attraction is temporary but it provides another example of the way adventure tourism is increasingly offered in urban contexts. For the companies involved, locating these installations in cities opens up larger markets – with demand from tourists and residential populations. For the same reasons, bungee jumping has become a predominantly urban phenomenon because it is more accessible to large numbers of people (Beedie 2005). This trend is also changing the geography of adventure attractions, with established adventure tourism operators like Zip World opening new facilities in urban locations. In 2015,

Figure 6.7: 110 Bishopsgate – Formerly Heron Tower, now called the Salesforce Tower (Photo: Tristan Luker).

Go Ape followed this trend by opening their first city centre site in London's Battersea Park. This company installs ladders, ropes, platforms and zip wires in trees creating an elevated playground. Go Ape in Battersea Park caused a lot of controversy because it meant an expensive attraction was installed in a public park. When a similar attraction opened in Glasgow, the urbanist Ronan Paddison (2010) was one of the people who campaigned against it – arguing that Go

Ape meant the privatisation of public space. In the era of neoliberalism, we are used to private incursions in the public realm, but Go Ape attractions are unusual examples because their installations are mainly above ground. In Battersea Park the playground beneath Go Ape remains free to use, but the playground in the trees costs £18–33 – creating a two tier park where: 'those who can afford it get to swing through the heavens and look down on those whose lack of cash leaves them scurrying about below' (Mangan 2015). Go Ape in Battersea Park means the commercialisation and privatisation of the vertical space in it; and it provides a further example of the way London's visitor economy is expanding vertically.

Out of View

Before concluding this chapter, it is important to mention some of the aspects of vertical city tourism that have not been discussed. The analysis here has consciously tried to focus on experiences from above – a perspective that ignores the importance of looking up – e.g. at tall buildings, suspended installations and airborne events. A ground floor perspective is not addressed but neither is an underground one – and this aspect is particularly relevant to London given its pioneering role in the construction of underground railways and river tunnels. As the discussion focuses on direct (i.e. non-representational) experiences, the significant use of panoramic views in marketing materials has also been neglected. Critics might also suggest the discussion has been overly positive – ignoring some of the darker aspects of towers and tourism. For example, urban towers and bridges have always provided opportunities for suicidal people wanting to end their lives. The tragedies at the World Trade Center in New York and, more recently, at Grenfell Tower in London, also highlight the potential for disaster that permeates tall buildings. These tragedies – both of which created disturbing icons which people wanted to visit – have not dulled the appetite for high rise urbanism. This suggests we have now entered an age where the growth of vertical urban space – and vertical tourism territory – is inevitable.

Conclusions: Urban Tourism in 3D

This chapter has provided a comprehensive overview of the new ways that tourists are now able to consume London from above. These include traditional observation decks installed in new skyscrapers (SkyGarden, The View from the Shard), viewing platforms incorporated into regenerated industrial structures (The Oxo Tower, Tate Modern) and moving attractions that simulate flight (The London Eye, The Emirates Airline). All these attractions have been opened in the last 20 years, with the Millennium celebrations and the London

2012 Olympic Games providing excuses to use public funds. The chapter has also reviewed a new breed of attractions which facilitate more physical experiences, including ways of climbing (Up at the O2), descending (The Slide, Zip World London) and traversing (Go Ape, The Tower Bridge Experience) the city. These experiences are based on the quirky appeal of consuming adventure tourism in an urban setting. Such attractions, alongside the plethora of bars, restaurants and hotels that have been opened in London's new high rise buildings, exemplify how London's visitor economy is expanding vertically. Over a century after New York developed elevated tourism and leisure spaces, London is following suit.

It is important to develop a critical understanding of the new vertical territories that are emerging in London, including an appreciation of the spatial politics through which socio-economic elites rise upwards (Graham and Hewitt 2012). These elites include tourists and the territories created are inextricably linked to tourism. Many of the attractions explored in this chapter have been co-produced by London's most lucrative industries – property, finance and tourism – which have combined forces to produce and commodify panoramic views of London. The tourism sector in London has been one of the beneficiaries of new high rise developments in the city, but it has also been one of the driving forces behind the city's vertical expansion. New office blocks are made more economically viable and socially justifiable by the introduction of viewing areas, hotels, and restaurants. These amenities generate rent, revenue and publicity and make high rise towers seem more public. London's new verticality has reinforced the increased socio-economic polarisation of the city and this is also linked to tourism. Affluent elites are keen to occupy central areas that are distanced vertically from the streets below and these elites include tourists who are keen to gaze on London from a range of exclusive vantage points without having to engage with the reality of this global city. Just as not everyone gets the chance to live or work in London's elevated territories, not everyone gets the chance to visit. Indeed, one of the recurring themes in the discussion here is the expense of the new attractions installed above London. These financial obstacles remind us that London's panoramic aspect is very much a privileged view and one that has been ruthlessly commodified.

The conclusions above suggest that many of the original explanations for the rise of elevated tourism attractions still apply in contemporary London. The popularity of new elevated viewpoints is based on the aesthetic appeal of urban panoramas and the attraction of being safely encapsulated high above the city. At one level these are innocuous attempts to access spectacular views, but they are also efforts to gain security in – and control of – the unruly city; thus, reinforcing the power dynamics of contemporary urbanism. The second half of the chapter explores the rise of different types of elevated experiences which are linked to the rise of the playful or ludic city (Stevens 2007), the desire for more active experiences, and the rise of urban adventure tourism. The observation deck has seemingly become a little old fashioned, and the ways in which this attraction has been reinvented (e.g. through the introduction of transparent floors, slides and virtual reality) demonstrates the rise of architecture designed to facilitate playful experiences (Deriu 2016). However, it would be a mistake to regard the new breed of high rise tourism as essentially different from traditional modes. Promotional materials produced by The Slide, Zip World London or Up at the O2 still emphasise that these are opportunities to consume London's panorama. The view is still very much central to their attractiveness. These attractions obviously involve exaggerated sensations of vertigo, but they are also used as vehicles to comprehend, control and own the city below. Visitors are still encapsulated by structures that separate them from the city, and whilst they involve more dynamic experiences, those consuming London's new breed of vertical attractions are distanced emotional and physically from urban reality.

References

Balm, Roger and Briavel Holcomb. 2003. 'Unlosing Lost Places: Image Making, Tourism and the Return to Terra Cognita'. In David Crouch and Nina Lübbren, eds, *Visual Culture and Tourism*. London: Bloomsbury.

Barthes, Roland. 1979. *The Eiffel Tower and Other Mythologies*. New York, NY: Hill and Wang.

Beard, Colin et al. 2003. *Adventure Tourism*. London: Routledge.

Beedie, Paul. 2005. 'The Adventure of Urban Tourism'. *Journal of Travel and Tourism Marketing*, 18(3), 37–48.

Borden, Iain. 2014. 'The Singapore Flyer: Experiencing Singaporean Modernity through Architecture, Motion and Bergson'. *The Journal of Architecture*, 19(6), 872–902.

Boyer, Christine. 1994. *The City of Collective Memory*. Cambridge, MA: MIT Press.

Brand, Peter and Julio Dávila. 2011. 'Mobility Innovation at the Urban Margins: Medellín's Metrocables'. *City*, 15(6), 647–661.

Clark, Greg. 2014. *The Making of a World City: London 1991 to 2021*. Chichester: John Wiley and Sons.

Cocks, Catherine. 2001. *Doing the Town: The Rise of Urban Tourism in the United States, 1850–1915*. Berkeley, CA: University of California Press.

Council on Tall Buildings and Urban Habitat (CTBUH). 2016. *Year in Review: Tall Trends of 2016*. Available at http://www.skyscrapercenter.com/year-in-review/2016 (accessed 7 December 2018).

De Certeau, Michel. 1984. *The Practice of Everyday Life*. Berkeley, CA: University of California Press.

Deriu, Davide. 2016. 'Don't Look Down!: A Short History of Rooftopping Photography'. *The Journal of Architecture*, 21(7), 1033–1061.

Deriu, Davide. 2018. 'Skywalking in the City: Glass Platforms and the Architecture of Vertigo'. *Emotion, Space and Society*, 28, 94–103.

Dorrian, Mark. 2009. 'The Aerial Image: Vertigo, Transparency and Miniaturi-zation'. *Parallax*, 15(4), 83–93.

Financial Times. 2014. *Hotel in London's Shard opens – with a Potentially Embarrassing Flaw*. Available at https://www.ft.com/content/d8564976-d687-11e3-907c-00144feabdc0 (accessed 7 December 2018).

Gilbert, David and Claire Hancock. 2006. 'New York City and the Transatlantic Imagination: French and English Tourism and the Spectacle of the Modern Metropolis, 1893–1939'. *Journal of Urban History*, 33(1), 77–107.

Gillespies. 2017. *Sky Garden visitor numbers go sky high*. Available at https://www.gillespies.co.uk/news/sky-garden-visitor-numbers-go-sky-high (accessed 7 December 2018).

Graham, Stephen and Lucy Hewitt. 2013. 'Getting off the Ground: On the Politics of Urban Verticality'. *Progress in Human Geography*, 37(1), 72–92.

Harris, Andrew. 2015. 'Vertical Urbanisms: Opening up Geographies of the Three-Dimensional City'. *Progress in Human Geography*, 39(5), 601–620.

Hinchcliffe, Tanis and Davide Deriu. 2010. 'Introduction Eyes over London: Re-Imagining the Metropolis in the Age of Aerial Vision'. *The London Journal*, 35(3), 221–224.

Jansson, André and Amanda Lagerkvist. 2009. 'The Future Gaze: City Panoramas as Politico–Emotive Geographies'. *Journal of Visual Culture*, 8(1), 25–53.

Leiper, Neil and Sun-Young Park. 2010. 'Skyscrapers' Influence on Cities' Roles as Tourist Destinations'. *Current Issues in Tourism*, 13(4), 333–349.

MacCannell, Dean. 1999. *The Tourist: A New Theory of the Leisure Class*. Berkeley, CA: University of California Press.

Mangan, Lucy. 2015. 'Why I'm going ape over the Privatisation of Children's Play'. *The Guardian*, 29 December. Available from https://www.theguardian.com/commentisfree/2015/dec/29/go-ape-privatisation-childrens-play-battersea (accessed 7 December 2018).

McKay, Tracey. 2013. 'Leaping into Urban Adventure: Orlando Bungee, Soweto, South Africa'. *African Journal for Physical Health Education, Recreation and Dance*, 19 (Supplement 3), 55–71.

NLA. (2017) *NLA Tall Buildings Survey*. Available at http://www.newlondon architecture.org/news/2017/march-2017/tall-buildings-survey-read-the-full-results (accessed 7 December 2018).

Olsen, Donald. 1986. *The City as a Work of Art: London – Paris – Vienna*. London: Yale University Press.

Paddison, Ronan. 2010. 'Protest in the Park: Preliminary Thoughts on the Silencing of Democratic Protest in the Neoliberal Age'. *Variant*, 39(40), 20–25.

The Shard. 2016. *A Record Breaking Year for the Shard*. Available at https://www.the-shard.com/news/2015-record-breaking-year-shard/(accessed 7 December 2018).

Smith, Andrew and Ingvild von Krogh Strand. 2011. 'Oslo's new Opera House: Cultural Flagship, Regeneration Tool or Destination Icon?'. *European Urban and Regional Studies*, 18(1), 93–110.

Stevens, Quentin. 2007. *The Ludic City: Exploring the Potential of Public Spaces*. London: Routledge.

Transport for London. 2017. *Emirates Air Line Performance*. Available at https://tfl.gov.uk/corporate/publications-and-reports/emirates-air-line-performance-data (accessed 7 December 2018).

Urry, John. 1990. *The Tourist Gaze. Leisure and Travel in Contemporary Societies*. London: Sage.

Wainwright, Oliver. 2015. 'London's Sky Garden: The More you Pay, the Worse the View', *The Guardian*. Available at https://www.theguardian.com/artanddesign/architecture-design-blog/2015/jan/06/londons-sky-garden-walkie-talkie-the-more-you-pay-the-worse-the-view (accessed 7 December 2018).

The Wharf. 2016. *Arcelor Mittal Orbit making Six Figure Sum after Opening of Slide*. Available at http://www.wharf.co.uk/news/local-news/arcelormittal-orbit-making-six-figure-1232607 (accessed 1 September 2017).

Wharton, Annabel. 2001. *Building the Cold War: Hilton International Hotels and Modern Architecture*. Chicago, IL: University of Chicago Press.

Zip World. 2017. *Zip World London*. Available at https://www.zipworld.co.uk/adventure/detail/zip-world-london(accessed 1 September 2018).

Acknowledgement

This Chapter was inspired by the work of Dr Davide Deriu and the Vertigo research project (see www.vertigointhecity.com for details). Vertigo in the City was an exploratory research project supported by a Wellcome Trust Medical Humanities grant and led by Dr Davide Deriu from the University of Westminster. The project team included three other academics: Dr Josephine Kane, Professor John Golding and Professor Brendan Walker. This six-month project culminated in a two-day event hosted by the University of Westminster (29-30 May 2015). This Chapter aimed to draw attention to this research project, and to apply the ideas it spawned to tourism and tourism development.

CHAPTER 7

London's 'Unseen Tours': Slumming or Social Tourism?

Claudia Dolezal and Jayni Gudka

Introduction

This chapter investigates the work of Unseen Tours, a not-for-profit social enterprise based in London which offers a source of income to homeless, formerly homeless and vulnerably housed Londoners by employing them as tour guides. The aim of this chapter is to raise awareness of the work of Unseen Tours in the context of London's changing visitor economy and to relate it to the wider debates on slum tourism (Freire-Medeiros 2013; Frenzel and Koens 2012) and societal change in the city (Paddison and McCann 2014). There is a fine line between selling and commodifying poverty and making a social contribution to poor peoples' lives, for example by creating new and alternative livelihoods as part of the ethical and responsible tourism agenda (OBrien 2011). This chapter calls not only for the inclusion of homeless tour guiding in the debates over the tourism-poverty nexus, but also for increased research efforts into this recent social phenomenon.

What follows is a discussion of the extent to which the tours could possibly be seen as a new kind of 'western' slum tourism, selling poverty as an attraction (Freire-Medeiros 2009), or whether they challenge prevalent perceptions

How to cite this book chapter:
Dolezal, C. and Gudka, J. 2019. London's 'Unseen Tours': Slumming or Social Tourism?. In: Smith, A. and Graham, A. (eds.) *Destination London: The Expansion of the Visitor Economy.* Pp. 141–163. London: University of Westminster Press. DOI: https://doi.org/10.16997/book35.g. License: CC-BY-NC-ND

of homelessness. Although the aims of slum tourism products vary, this kind of tourism has faced major criticism in recent years, with commentators questioning the ethics of tourism consumption based on poverty (Meschkank 2011; Rolfes 2010; Freire-Medeiros 2013). By discussing the relevant literature, as well as reflecting on the work of Unseen Tours, and comparing it to international examples of homeless tour guiding, this chapter argues that the project has the potential to contribute to positive social change in line with the ideals of social tourism (McCabe et al. 2012). While this chapter acknowledges that Unseen Tours cannot solve the homelessness problem in London, the organisation does have the power to create new opportunities and visibility for those experiencing homelessness whilst enabling them to play a role in in London's tourism sector. At the same time, the tours fulfil tourists' ever-present demands for encountering the 'authentic' (see Chapters 2 and 3) and help to diversify the tourism offer in London, distributing benefits beyond the traditional tourist centres to more 'edgy' urban destinations (Smith and Pappalepore, 2015), in line with the territorial expansion of tourism in the city.

'Unseen Tours': The History of a Social Enterprise in Tourism

Unseen Tours is a London-based social enterprise that was founded in 2010 to address the rising problem of homelessness in the UK's capital city. In a country as prosperous as the UK, the growing numbers on homelessness remain shocking, with government statistics revealing that in the year 2017, 4,751 people were sleeping rough in England's streets, which constitutes an increase of 15 per cent from the previous year (Homeless Link 2018). London is at the very centre of the problem. Out of all regions in England, London alone saw 1,137 rough sleepers in 2017 (Homeless Link 2018). These numbers do not even fully reflect reality as they do not include the 'hidden' homeless, with people couchsurfing on friends' or strangers' sofas or sleeping on public transport (Butler 2018). Some argue that the Conservative government's policies are to blame for the situation in London, including a lack of supply of affordable housing, and a cut in housing benefits as well as funding for homelessness services (Butler 2018). More specifically, the Local Housing Allowance Reforms that started in 2011 introduced obstacles preventing lower income households from accessing private tenancies, particularly in inner London (Crisis 2018). While the government is now trying to take action to address this problem, for example with the coming into force of the Homelessness Reduction Act in April 2018, key problems remain which 'relate to the growing structural difficulties that many local authorities face in securing affordable housing for their homeless applicants' (Crisis 2018, xiii). In addition, the social stigma around homelessness is a major problem, with a study by homelessness charity 'Evolve' revealing that 72 per cent of people in the UK believe that homeless people themselves are to blame for being in or remaining in the streets (Evolve 2018).

Unseen Tours seeks to address the problem of homelessness by employing homeless or vulnerably housed Londoners and raising awareness to put this problem back on the government's priority list. The organisation was established as part of the 'Sock Mob', a small group of people who ventured into the streets of London with food, drinks, material goods (including, but not limited to, socks), to listen to, to talk and to learn from London's rough sleepers. Since it began, The Sock Mob has grown into a more than 600-strong network of like-minded individuals from all walks of life, providing companionship and support for lonely and isolated people living on the streets. The network enables people to socialise and meet in different social contexts, including boat trips, bowling, picnics and other leisure activities. The idea of walking tours led by homeless people emerged from these activities and evolved into a formal enterprise: Unseen Tours.

Unseen Tour's members were inspired to create the very first pilot program after gaining a closer understanding of homelessness and seeing how creative and resilient their street friends were. They found that homeless people had great stories to tell, but often did not have access to paid opportunities to harness and nurture these qualities. Their intention was to come up with a different and revolutionary idea that would change perceptions of homelessness by cutting through negative stereotypes and social stigmas. The idea that gave birth to the organisation was twofold: changing how people saw and thought about homelessness, whilst helping vulnerable individuals directly. The social enterprise officially launched its first tours in August 2010, and since then it has employed 20 homeless and formerly homeless tour guides who have guided both national and international tourists around different areas of London: Brick Lane, Camden, Shoreditch, London Bridge, Brixton, Mayfair and Covent Garden. This list includes areas that are off the beaten track but also those that are normally part of tourists' bucket lists. Unseen Tours still has a close relationship with the 'Sock Mob', and both organisations continue to support homeless people and challenge attitudes towards homelessness through their own projects in London.

London is not the only European city where homeless tour guiding allows tourists to explore the city off the beaten track. Vienna, Prague, Berlin and Barcelona also offer organised tours by homeless and vulnerably housed residents. For example, 'Shades Tours' in Vienna (established 2016) aims to reintegrate some of the 10,000 homeless people into the Austrian job market. It currently employs 18 guides and has already helped two to secure their own flats, and another two to enter the job market (Shades Tours 2018). According to Shades Tours, the aim of the project is empowering homeless people by giving them a task and motivation, self-confidence, an income and new opportunities (Shades Tours 2018) whilst helping visitors change their views of homelessness, thus increasing tolerance more broadly.

Changing stereotypes is also at the core of the homeless tour guiding project in Berlin, where 'Querstadtein' (established 2013) has bridged the divide between 'homeless people and the rest of society ... [and] create[d] a space which would facilitate encounters, exchange and awareness' (Querstadtein 2016a, n.p.). One of the tours on offer is 'biographical ... [and] touches on East–West German history, survival strategies for both the past and present, and the designs and uses of public spaces' (Querstadtein 2016b: n.p.). More recently, the project also started to involve refugees as guides to increase understanding of the challenges faced by this vulnerable group and to foster integration in times of crisis (Querstadtein 2016b). Querstadtein is a good example of using tourism as a tool for empowering disadvantaged residents while, at the same time, addressing the issue of a different kind of mobility – migration (UNWTO 2009).

An in-depth analysis of the above-mentioned projects lies beyond the scope of this chapter, but important differences can be noticed. Not all of the homeless tour guiding initiatives currently recorded in Europe take the form of not-for-profit social enterprises, which means that, while achieving social impact, some are essentially profit-orientated. The motivation behind Unseen Tours is not profit-making but philanthropy. Its members can be seen as 'change agents' as Sharis and Lerner (2006) regard many of today's social entrepreneurs who aim to achieve social and environmental progress. Some subsume contributions that social enterprises make under the broader terms of 'community tourism' or 'responsible tourism' (Mottiar 2016). These normative ideas have shaped tourism literature and practice for several decades and aim to achieve sustainable benefits for local communities and environments (Goodwin 2011; Scheyvens 2002; Telfer and Sharpley 2015). The important difference is that social enterprises are characterised by specific business models which generate profits just like traditional businesses but reinvest these profits into social and environmental causes (Bornstein and Davis 2010). Hence, as Sheldon et al. (2017) argue, a 'social entrepreneur can simply be defined as one who uses business principles to solve social problems' (4). Social entrepreneurship is therefore a more 'modern' way of pursuing philanthropy by initiating projects that are proactive and lead to sustainable change, rather than focusing on reactive giving through donations and Corporate Social Responsibility (CSR) (Novelli et al. 2015).

As a social enterprise working in tourism, Unseen Tours' not-for-profit business model is unique in that the organisation is run by unpaid volunteers who have full-time jobs, and work on Unseen Tours in their spare time. They support tour guides but it is the guides who design the tours and their content. This is not the case in other homeless tour guiding projects in Europe. At least 60 per cent of the money raised from the sale of Unseen Tour tickets goes directly to the homeless and vulnerably housed tour guides, with the remaining revenue used for essential operational expenses. This allows the social enterprise to cover guides' expenses (mobile phones, transportation costs), training and upskilling, as well as any essential operational costs. Any profits Unseen Tours make are directly reinvested back into the enterprise to engage more guides and widen the scope of their tours. In doing so they 'balance social goals with the need to generate revenues' (Day and Mody 2017, 67), a challenge that most social entrepreneurs face. The volunteers who run the organisation do not financially profit from the organisation, and all the money raised through ticket sales is used for the social purposes for which Unseen Tours was founded. Unseen Tours also works to challenge the negative stereotypes associated with homelessness more broadly through their social media, outreach, events and the content of the tours. This remains an important part of the organisation's work to this day, also for the guides themselves:

> I like it because you can change people's perceptions on homelessness, and [they] see us as individuals rather than a group. They get a better understanding of what it's like to sleep rough and how they struggle to get by, how lonely it can be and that it can easily happen to anyone (Viv, Covent Garden guide).

Unseen Tours assists in 'includ[ing] groups into tourism that would otherwise be excluded from it' (Minnaert 2014, 283), which is essentially how social tourism is defined. This inclusion can either be of disadvantaged tourists by enabling them to participate in tourism activities, or of residents who are excluded from the tourism industry. Minnaert et al. (2006) discuss visitor- and host-related social tourism, with the former focusing on Western countries while the latter concerns mainly residents in developing countries.

Figure 7.1: Viv on a Tour through Covent Garden (Photo: Unseen Tours).

Unseen Tours operates within the Global North, but many of the social enterprises focusing on the inclusion of disadvantaged residents in the tourism industry can be found in developing countries. They are often connected to alternative tourism development projects, which should offer more ethical, responsible and locally beneficial alternatives to mass tourism (Mowforth and Munt 2016). While these developments (such as community-based tourism or ecotourism) are centred on a participatory and community empowerment ethos (Scheyvens 2002), they also constitute a more morally sound product for consumers (Butcher 2003), and satisfy tourists' search for authenticity (Dolezal 2011; 2015a). Butcher (2003) questions the raison d'être of these alternative forms of tourism. He suggests these fulfil the desires of tourists, in line with wide-spread neoliberal ideologies, whilst failing to deliver tangible change on the ground.

Poverty has become an attraction in itself (Baptista 2010; Freire-Medeiros 2009), particularly in the developing world, where slum, orphan, poverty or charity tourism increasingly appear on tourists' itineraries. This commodification of poverty is regarded as problematic not only because of the voyeurism involved, but also for benefiting an elite that uses poverty for their own ends (Frenzel 2013). Homeless tour guiding is not exempt from this kind of criticism. *The Guardian*, for example, refers to Prague's social enterprise Pragulic as a kind of 'poverty tourism' (Allen 2016) and *The Independent* discusses Unseen Tours' work under the terms of 'poorism' and 'slum tourism' in a Western context (Taylor 2011).

A cursory glance at the idea of homeless tour guiding may evoke certain similarities to slum tourism and poorism, by turning poverty into an attraction in Western cities. The question is: are such parallels justified or does the work of the social enterprises discussed above, particularly that of Unseen Tours, contribute to positive societal change? This chapter will try to answer this question and point to the ways that Unseen Tours enables London's marginalised residents to empower themselves.

The Tours: London's Vulnerable Residents at Work

A typical Unseen Tour starts by meeting a guide in the chosen area who will introduce themselves and explain what led them to join as a guide. Tourists then follow the guide on a walking tour through the neighbourhood, which is usually also the area where the guide him/herself lives. According to Frenzel and Blakeman (2015) this ultimately signifies to the tourists that the knowledge the guide transmits is more credible and 'real', thus making the tour overall more 'authentic' (this notion will be discussed later in the chapter). The content of the tours varies from person to person, but usually the guides share historical and social knowledge of the area, intertwined with their own opinions and experiences of homelessness. The areas of interest include major attractions and some lesser known and personal ones which may not have previously been evident as attractions to tourists. On Trip Advisor, tourists speak mainly of the educational aspect of these tours: they appreciate seeing a new area of London, but also like gaining new knowledge about it.

Finding original paintings by the famous street artists 'Stik' or 'Banksy' is not something that tourists necessarily expect from a homeless tour in London's Shoreditch, for example. Other tours bring visitors to hidden community gardens on the Brick Lane Tour or Bridget Jones' front door near London Bridge – thus helping to uncover some more hidden elements in the city that appeal to tourists. The same is the case when tourists learn more about the meaning that the guides attribute to certain places in the neighbourhood, the best and warmest places to find shelter and hear some personal stories about daily experiences of homelessness and social stigma. Figure 7.2, for example, portrays David, the London Bridge tour guide showing tourists the Shakespeare street art near the Globe Theatre, while also talking about the anti-homeless architecture in the area, which was constructed to prevent homeless people sleeping in public spaces. The guides thus narrate their own stories and experiences throughout the tours which gives tourists a different perspective on the city and makes every tour unique. The purpose of the tours is to show people a historical London with an 'unseen' dimension, through the lens of homelessness.

Another aspect of the tours is the conversations they create about other social issues that the guides are passionate about. For example, the Brick Lane tour

Figure 7.2: David on the London Bridge Tour Discussing Homelessness and Social Stigma (Photo: Unseen Tours).

looks at the history of suffrage and multi-faith communities in London, whilst the Shoreditch tour highlights gentrification, which is extremely visible in that part of London. This opportunity for the guides to weave in their personal opinions with local and world history allows visitors to see London in a new light. Viv, Unseen Tours' Covent Garden guide, confirms this:

> I enjoy looking for unusual stories and teaching people about London's history, meeting people from different countries and walks of life. I just love doing tours. I think my tours help people see homelessness differently as I can tell people about my own experience. As my route is set along where I used to sleep rough, half of the tour focuses on that, and the other half is local history and interesting stories. Also I talk about homeless day centres and other things to do with homelessness.

Like Viv, any vulnerably housed Londoner can become a guide for Unseen Tours. In fact, many of the guides are recommended by other, already established, guides. When the pilot tours started in 2010, only three guides were involved. By 2018, the organisation was working with 20 guides. The areas that are being visited are the ones that the guides know best, and the content is based on facts and issues they are passionate about. Often guides have a location or route in mind which they would like to cover. While the guide is free to talk about their own experiences and stories, this does not mean that no training is provided.

After an initial meeting to see whether the guide would be a good fit for Unseen Tours and vice versa (i.e. whether the organisation could usefully assist the individual), regular meetings take place between the guide and a buddy who explore the area together, usually once a week over a few months. Through the research that the guide undertakes and the conversations and walks they have with their volunteer buddy from Unseen Tours, ten to twelve major points of interest are selected for the main route of the tour. This makes the training rather informal, facilitated through volunteers and experienced guides. The tours are therefore co-constructed by guides and buddies alike, with the guides' own voices and interests taking priority, with Unseen Tours acting as a mentor and facilitator to draw out their ideas and strengths. Some of the volunteers have experience and backgrounds in acting and tour guiding, which is particularly useful if the guide needs help with the 'performance' element of tour guiding. In such cases, the volunteers support guides with additional coaching on public speaking, presentation, projection, and more. More experienced guides also help with this through informal meetings as the tours develop. Ultimately, the whole exercise is designed to boost the guides' confidence and help them realise their own potential, which is why Unseen Tours is offering this kind of support. The duration of this part of the mentoring process varies largely as it depends on the guides' own readiness, circumstances and reliability.

Once the guide is confident of the route and feels ready to launch, Unseen Tours schedules at least two dress rehearsals, inviting all the volunteers and other Unseen Tours guides so that everyone can get to know the new guide and route. Unseen Tours also gets in contact with the guide to see if they need anything, including new clothes, shoes or similar equipment and gets it for them before the launch of a new tour. The ticket sales help to pay for this.

In terms of a long-term vision for the guides' future, the social enterprise sees no one-size-fits-all solution. While some of the other social enterprises discussed above aim to get homeless people back into working life as quickly as possible, Unseen Tours works with each guide on an individual basis. It therefore prefers to find solutions tailored to individual needs. In addition to offering employment, Unseen Tours supports guides seeking new or better housing arrangements and those going through difficult periods. Some previous guides have gone on to different kinds of work after their tour guiding experience, both within and outside of the tourism sector.

As the above discussion shows, Unseen Tours has worked hard to empower their guides, such as, for example Viv, who says 'I love being a part of it. It gives me a purpose.' Besides empowering guides, the organisation also ensures that homeless people are not exploited or disrespected through their tours in a voyeuristic way. Unseen Tours' vision of social inclusion and positive change is highly commended by a range of sources (Bland 2015; Pati 2010; Trip Advisor), and the enterprise won the Responsible Tourism Award in 2011 as well as the travel category at the 2013 Observer Ethical Awards. However, others argue that the social enterprise facilitates poverty or slum tourism (Taylor 2011) by turning poverty into something that can be consumed by tourists. Therefore, it is worth analysing whether any real parallels exist between homeless tour guiding and slum tourism.

Homeless Tour Guiding in the Context of Slum and Poverty Tourism

Traditionally, slum tourism is known as a phenomenon mainly to be found in metropoles in developing or emerging countries, including Brazil, India and South Africa (Rolfes 2010). However, its origins can be traced back to Europe where 'slumming' was a popular leisure activity for the upper and middle class in London, who visited poorer quarters of the city at the end of the nineteenth century (Koven 2006). Today, slum tourism is an organised industry with increasing numbers of businesses offering slum tours in poorer quarters of the Global South (Meschkank 2011), particularly for the international tourist market (OBrien 2011). This kind of tourism has been subjected to strong criticisms for many years, in that it is often regarded as selling poverty and stimulating voyeuristic tourism, which is why it has also been referred to as 'poverty tourism', 'human safaris', 'poorism' or 'negative sightseeing' (Freire-Medeiros 2013; Meschkank 2011).

Although a 'typical' slum tour is hard to define, people often think of them as being voyeuristic and exploitative, where visitors romanticise poverty in segregated, unfamiliar and inaccessible spaces. A range of researchers have investigated the motives that attract tourists to slums and the role that poverty plays (Rolfes 2010; Meschkank 2011; Burgold and Rolfes 2013). There is a general consensus that tourists embark on a slum tour to fulfill their desire to experience 'authentic' or 'real' aspects of the country they are visiting, which they hope to find in its poorest areas, such as slums, townships or favelas (Rolfes 2010; Meschkank 2011). Indeed, the search for authenticity while travelling is not new – after all, MacCannell (1999) argues that the tourist's main characteristic is the restless search to get beyond what is obviously presented in order to see the hidden and real destination (although failing to do so). The kind of 'authenticity' tourists often experience is staged culture performed in touristic spaces, rather than 'real' life in the destinations they visit (MacCannell 1999). One may argue that the reason for this ever-increasing search for authenticity is because we find ourselves in a highly globalised world, characterised by modernity and a loss of meaning and real human relationships (Bauman 2010).

'Off the beaten track' alternatives to mass tourism, such as ecotourism and community-based tourism (CBT) have emerged in response to tourists' desire to come closer to residents – in addition to the need to increase benefits for residents in destinations. As part of an earlier piece of research, one of the authors found that authenticity was a key selling point of the CBT produc but, most importantly, that it was related to the 'underdeveloped charm' of villages (Dolezal 2015a). In a study on edgy urban destinations in London, two of the other authors that appear in this book, Smith and Pappalepore (2015), discovered that authenticity for tourists often relates to more edgy areas in the city, which are 'chain free', characterised by ethnic diversity and where one can meet 'real Londoners'. Similar to these examples, in the context of slum tourism, tourists tend to connect the notion of authenticity directly with ideas of poverty – as demonstrated by Meschkank's (2011) observations on slum tourism in India:

> For the tourist [...] the real in the sense of authentic India is the poor India. Thus, a relationship can be identified between the degree of authenticity and the grade of poverty. That tourists look for authenticity off the beaten track is not new, nor is the notion that this authenticity increases with the grade of poverty. For at least as long as the Lonely Planet handbook—the traveller's bible, has marketed tours off the beaten track, these tours have gone to the poorer countries in the world. However, it is new that these tours go to slums (53).

Meschkank makes the argument that poverty has always stimulated interest amongst tourists, particularly when they themselves live in more industrialised nations than those they visit. However, the role that poverty plays in the touristic product becomes problematic when poverty turns into a commodity fetish, which is romanticised and traded for money (Frenzel 2013; Selinger 2009). Freire-Medeiros (2009) goes as far as speaking of the 'exchange value' of poverty in the sphere of tourism:

> [A]lthough under capitalism every single thing may be turned into a commodity, [Marx states that] there is one thing which can never be bought or sold: poverty, for it has no exchange value. The fact is that at the turn of the millennium, poverty has been framed as a product for consumption through tourism on a global scale (586).

If there is a danger of commodifying poverty as part of slum tourism (Frenzel 2013) or CBT (Dolezal 2015a), the question that emerges is whether homeless tour guiding uses a similar kind of attraction through the interaction with their tour guides? This raises a further question: if poverty is part of the appeal, what role can tourism really play in its alleviation? If poverty was the main attraction, this would mean that a lasting positive change for residents and destinations would at the same time destroy the very attraction that tourism depends on. Therefore, it is important to understand what really constitutes the attraction in different kinds of tourism. Slum tourism, poverty tourism and homeless tour guiding deserve closer analysis given the obvious role that poverty plays. Notably in the context of slum tourism, research has been conducted into the motivations of tourists, revealing that poverty often only forms the background context of the tours, with tourists primarily having an interest in residents' 'real' life, rather than poverty per se (Rolfes 2010). More importantly, the general meaning of poverty and the words associated with the slum were found to change after the tours and developed into more positive ones, including ideas of 'friendliness' and 'community' and an overall organised system in the slum (Dyson 2012; Meschkank 2011; Rolfes 2010). The tours therefore have the potential to relativise the idea of poverty and change tourists' preconceptions of poverty and poor urban areas (Monroe and Bishop, 2016).

Frenzel (2012) points towards the power that slum tourism bears for change, by turning poverty into something of importance in the context of tourism:

> The transformation of the slum into something valuable fulfils another, moral, function for the tourist. If poverty is understood as a problem, gazing at it evokes the necessity to do something about it. Indeed, voyeurism means that poverty is simply consumed for entertainment with no regards for the poor; they are simple 'othered' as poor. However,

> if poverty can be seen as something valuable, touring it becomes a 'must-do' just like other valuable sights need to be 'seen'. Arguably this is what has happened in the most developed slum tourism destinations. 'Othering' is recast in a positive light: the poor are 'others', but they are good! Concurrently there is less need to do something about their poverty (59).

Hence, while the 'othering' of poor residents can lead to objectification through voyeuristic encounters (Selinger 2009), it does not necessarily have to be characterised by power inequalities (Dolezal 2015b), but instead can lead to increased awareness about pressing issues (Frenzel 2013). Indeed, the benefits that slum tourism generates for residents have been subject to a certain amount of research (Dyson 2012; Freire-Medeiros 2012; Mekawy 2012), also by NGOs (Monroe and Bishop, 2016). Tourism Concern, a UK-based NGO working on tourism and human rights issues, emphasises a range of benefits, such as socio-economic empowerment and pride in one's community and life (Monroe and Bishop 2016) that such tourism can bring. However, there is a need to better understand how residents are involved in slum tourism decision-making processes and what their position is (Frenzel and Koens 2012). To date, little research has been conducted on residents' views of slum tourism (Freire Medeiros 2012), as is the case with many other kinds of tourism. The success and benefits of slum tourism for local residents largely depend on how tours are organised, i.e. in regard to representation but also economic and political aspects (Dürr and Jaffe 2012). When managed and organised well, some even talk about 'responsible slum tourism', which is in line with pro-poor tourism's general aim of benefiting the poor and creating economic linkages (Mekawy 2012).

Questions that need to be asked in order to better understand opportunities for empowerment in tourism relate to where the economic benefit of tourism goes, whether tourism supports wider community projects in the area or benefits only individuals and in how far residents can represent themselves as part of tourism (Dolezal 2015a). In regards to the latter, the role of the guide is important here, in that they are mediators and cultural brokers, who interpret and therefore shape reality for the tourists (Hallin and Dobers 2012; Salazar 2005). They endow space with meaning, depending on their stories and interaction with the space (Hallin and Dobers 2012). Through their narratives, they can create attractions (Frenzel and Blakeman 2015) and can possibly, as is the case with homeless tour guiding, become attractions themselves. At the same time, if representation is in the hands of external guides rather than residents, this 'can sometimes package poverty in exotified and romanticised ways, creating false tourist perceptions of real life in slums. This also takes away the agency of people who actually live in these areas to present their own personal narratives' (Monroe and Bishop 2016, 6) and, as a consequence, leads to their disempowerment (Dolezal 2015a).

In this context, it is the political role of slum tourism and tour guiding in general, that deserves much more research attention. After all, as Hallin and Dobers (2012) state, 'the guided tour, regardless of its intention, is political in nature'(23). Frenzel (2014) even goes so far as to argue that slum tourism forms part of wider urban regeneration from below, given that it 'responds to an absence of action or perceived failure to respond to poverty by urban policy' (431). Tourism, thus, has the power to acknowledge communities marginalised by the government and put them back on the map (Frenzel 2014).

In the context of homeless tour guiding, the ideas above relate closely to the work of Unseen Tours, particularly in respect to guide's empowerment and the changing of social stigmas. The range of counter-arguments to the criticisms of slum or poverty tourism that already exist serves as a basis to develop our ideas on homeless tour guiding and its yet under-acknowledged role in urban regeneration and social change in the city.

Empowering Vulnerably Housed Londoners through Tourism

Homeless tour guiding is a more recent phenomenon than slum tourism, and little research has been conducted to date. Only a small number of authors have referred to homeless tour guiding projects in the context of wider research on social enterprises (Dredge 2017; Kraftova et al. 2015). However, so far, no in-depth analysis of the phenomenon itself, tourists' motivations or guides' views has beenconducted. In fact, much of the writing on Unseen Tours to date can be found in the media, such as the BBC (Jarosz 2012), the *Independent* (Taylor 2011), the *Telegraph* (Morris 2015) or the *Guardian* (Bland 2015). As stated earlier, opinions vary on the topic, with some authors criticising the idea of homeless tour guiding and others celebrating the social enterprise's successes in empowering guides and challenging stereotypes.

As a result of this lack of research, this chapter's final section is an attempt to bring together a range of key arguments that respond to some of the criticisms of homeless tour guiding, pointing towards the positive change it can bring. It does so by drawing on the relevant literature on slum tourism, tour guiding and empowerment and combining it with the personal reflections and experiences of the authors.

The Gaze is not Directed at the Exotic or Economically Poor

Homeless tour guiding has experienced criticisms in online newspapers and on travel websites for selling poverty as a commodity – evoking parallels with slum tourism and poorism (Kassam 2013; Taylor 2011). As discussed above, these kinds of tourism are often chastised for directing the tourist gaze onto the economically disadvantaged exotic, resulting in voyeurism and unequal power relations between tourists and residents. While this may be the case in slum tours to a certain extent, Unseen Tours actively works to avoid to 'voyeuristically and superficially [point] out economically and "socially" deprived areas and the people within them' (Unseen Tours 2016, n.p.). Guides are empowered to share their own personal stories rather than pointing to and sharing those of others. Figure 7.3, for example, shows Mike on a tour in Camden, telling his own stories about the area as well as his own experiences of homelessness.

This does not mean that Unseen Tours ignores the differences in wealth that may exist between tourists and residents. As Dürr and Jaffe (2012) remind us, once tourism involves pronounced class differences between 'host' and 'guest', one needs to consider the ethical aspects of it. The 'othering' of guides may happen in a sense that guides are indeed different from the tourists due to their economic and living conditions. However, as a number of authors have shown elsewhere in the context of CBT and slum tourism (Dolezal 2015b; Frenzel 2013), the othering of residents can also create an awareness of difference, poverty and other social problems – ultimately leading to greater mutual understanding. The Unseen Tours guides' position of authority (as discussed below) tips the unequal power relations of tourism on its head, as the subjects of the tours become authoritative figures leading tourists around their areas. Thus, through Unseen Tours, the tourist gaze is redirected into different spheres, away from the obvious features of poverty and the traditional attractions in London, towards the unseen elements of the city and the realities of homelessness.

Figure 7.3: Mike on a Tour in Camden (Photo: Unseen Tours).

Poverty is not the Key Attraction in Homeless Tour Guiding

While the role that poverty plays in slum tourism is debatable, the discussed research has confirmed that a certain element of poverty, often disguised as 'authenticity' for tourists, forms part of the attraction. When it comes to homeless tour guiding, poverty in the form of the economic disadvantage that most homeless and vulnerably housed people experience undoubtedly plays a role; however, it is not an attraction in itself. Unseen Tours' intentions are to avoid exploiting guides for financial benefit, which is also why the organisation was established as a not-for-profit social enterprise. The customers of Unseen Tours do not come to see poverty in London – instead, they come to learn more about the city's past, local stories and see the city through the perspective of local experts. This does not mean that the guide is not an attraction him/herself: having a homeless tour guide is often a fascinating experience for visitors.

Considering the key role that tour guides play for tourists in interpreting and mediating spaces (Reisinger and Steiner 2006), one may argue that they don't just create attractions (Frenzel and Blakeman 2015) but can become an attraction in themselves. It is the knowledge and performative narrative of the guide that constitutes the attraction of the tours rather than the guide's vulnerable housing situation. Indeed, 'performance' is a key element of the tours – as is the case with tour guiding in general (Jonasson and Scherle 2012). The guide therefore assumes a double role – while becoming an interpreter/broker in the touristic context, their life is, at the same time, part of the content of the tour and the daily reality of the toured space.

Through training and coaching provided by volunteers, the guide has a chance to articulate his/her agency on this stage and can choose how best to present him/herself and the content of the tours. This gives him/her a certain power to decide on what s/he sells as attraction as part of the tours. It ensures that the power dynamics of this tourism system are not just one-sided (Cheong and Miller 2000), and the attraction itself is co-constructed in the touristic process, i.e. the tour itself. The empowerment of residents depends largely on what the attraction constitutes (Dolezal 2015a). Therefore, it is important to understand tourists' motivations to be part of the tours as well as the guides' role therein.

Homeless Tour Guiding does not Depend on Segregated City Environments or the Continuation of Homelessness

Slum tours usually lead tourists to deprived and poor areas of urban environments, but homeless tour guiding does not depend on visits to segregated areas. Besides less popular tourist areas, Unseen Tours' guides also work in established tourist areas, found on the 'typical' tourist trail, including Camden, London Bridge and Brick Lane. These are already familiar to many of the visitors on the tours, which is why Unseen Tours shines a light on how, in many instances, situations of deprivation and social exclusion are hidden behind the 'glitzy' touristy facade. This means that the tourist gaze is redirected to a different sphere. Visitors are looking to learn more about the areas from a different perspective and, most importantly, from a local resident who is from the area, which makes touristy spaces more authentic in the eye of the visitor. (Frenzel and Blakeman 2015). However, this does not automatically mean that authenticity relates to poverty in this case – an aspect that deserves further exploration.

The way that Unseen Tours operates means that their guides do not need to remain homeless or vulnerably housed for them to remain a tour guide with the organisation. Though there has been criticism of poverty tourism projects in this regard, in which it has been argued that the poverty or deprivation which constitutes the subject of the tour is required to remain constant and unchanging for the tours to continue, this is not the case here. In fact, the Unseen Tours volunteers work with the tour guides to help them find temporary and permanent housing and provide continuous support through their buddy schemes. The volunteers are keen to emphasise that they would be happier if homelessness would cease to exist, and their continued employment does not rely on their housing status.

Homeless Tour Guiding Can Empower Guides

This chapter argues that homeless tours empower guides, giving them a platform and voice to tell their stories and paint a picture of the city from their own perspective. It is an opportunity to add alternative narratives to the dominant tourism discourses of the city, which tend to overlook social inequalities and poverty. Unseen Tours ensures that 60 per cent of their sales go directly to guides, and the remaining money is spent on training, marketing and other essentials, such as mobile phone tariffs and transportation for the guides. In contrast to other homeless tour guiding companies, Unseen Tours is entirely run by volunteers who support guides with coaching and training, which is another key aspect when it comes to empowerment in the context of tourism (Dolezal 2015b; Scheyvens 2003). Indeed, the tours are very much designed to elevate the guides to a position of authority where the customers pay to come to listen to them share their knowledge – rather than simply giving them money. Guides are thus in a position to undertake their own tours and receive feedback which can, in turn, have an impact on their overall self-confidence, their sense of belonging to the area and the city. For example, David, one of the guides working for Unseen Tours, commented that, 'I am now doing a job which I love, I tell people that I used to go to work because the government told me to, now I go to work because I enjoy it!' Working as an Unseen Tours Guide therefore gives guides a purpose and, ultimately, can also increase their ability to take up other work opportunities in the future: 'If I should ever leave Unseen

Tours I would get some qualifications and try to find more work as a tour guide or something similar. I would then have all three: qualifications, experience and skills. With those I should be able to get myself a job.'

From the very start, the main aim of Unseen Tours was that every aspect of the tours was co-constructed with the guides, empowering them to shape the narrative and how the tours would work. This is reflected in the tours them-selves and in how they are set up. Every effort is also made to make Unseen Tours a structurally and hierarchically flat organisation. Although the social enterprise has Directors in a legal sense, everyone involved in Unseen Tours has a say as to how the organisation is run and how to drive it forward. It is the guides that have the most say in the organisation, and they are at the core of its daily operations. Kinder (2016) confirmed this in earlier exploratory research on homeless tour guiding, in which she found that Unseen Tours encourages a sense of ownership and participation as much as economic empowerment.

Homeless Tour Guiding Challenges Social Stigmas about Homelessness

A review of the literature on slum tourism has shown that tourists' views of poverty and slums often change after having participated in a slum tour (Dyson 2012; Meschkank 2011; Rolfes 2010). It is well-known that tour guiding carries a strong educational role; however, it was found that particularly more 'alter-native tours' seek to educate on social issues and change (Byron 2012). Byron (2012) argues that these more alternative tour guides, which are not part of commercial organisations:

> are very engaged within their communities and often have an educa-tional mission. Making people think of social topics such as diversity or heritage is important; this is why tours are often described as a 'crosso-ver of work in socio-cultural education and tourism'. [...] [I]n contrast to official guides there is an additional motivation for tour guiding: they want to emancipate tourists, affect changes in tourists' personal lives and teach them to look beyond the alleged 'traditional tourist story' (34).

Problems of income polarisation and workers earning less than the London Living Wage often remain below the surface in elite-driven public discourse, creating stigmatisation rather than responses to poverty (DeVerteuil 2014). Homeless tour guiding can be a way of putting vulnerably housed Londoners 'back on the map' and giving them political power by being responsible for part of the tourism industry in the city. While the tours may not solve the problem of homelessness in London, they can help challenge the stereotypes associated with homeless people. Guides are empowered to showcase their unique experi-ences, knowledge, and skills and participate actively in the creation of value for incoming tourists.

Homeless Tour Guiding brings Wider Benefits for Urban Destinations

While no research has investigated the role of homeless tour guiding from a destination perspective to date, this must not be overlooked. Homeless tour guiding does not happen necessarily in the most popular tourist areas in London. Although some of the tours take tourists to familiar and popular tourist places and show these areas in alternative ways, they also cover peripheral areas that are increasingly interesting for tourists. They assist in dispersing tourists away from the traditional sightseeing paths between Big Ben and the South Bank, for instance, which can help overcome overcrowding and spread the economic benefits of tourism to different areas of London. The phenomenon of homeless tour guiding thus potentially contributes to the growth of tourism in more 'ordinary', edgy and impoverished urban neighbourhoods (Smith and Pappalepore 2015). As some of the tours bring tourists to more deprived areas of London they can serve as an 'urban development and regeneration tool from below' (Frenzel 2014, 431). At the same time, homeless tour guiding not only makes tourism more inclusive, but also offers a more experiential tourism product which uses storytelling and alternative narratives to satisfy the desires of the postmodern tourist (Byron 2012).

Conclusions

In this chapter, we have discussed the work of Unseen Tours and the phenomenon of homeless tour guiding more generally. The discussion outlined some of the criticisms that have been levelled at homeless tour guiding experiences, particularly when it comes to the commodification of poverty and the 'othering' of poor and vulnerably housed tour guides. While this criticism deserves further empirical investigation, this chapter argues that homeless tour guiding is driven by tourists' desire to learn about the 'real' and 'authentic' London. Homeless tour guiding takes tourists off the beaten track to discover novel attractions. Whilst authenticity is not necessarily related to notions of poverty, meaning that the authentic London is not necessarily the poor London (as is often the case with the areas visited in slum tourism), poverty does have a role to play as Unseen Tours' guides have had some experience of homelessness, vulnerable housing and economic disadvantage.

Therefore, while certain parallels with slum tourism may exist, the argument that homeless tour guiding is the new form of slum tourism or poorism seems difficult to justify. The locations where the tours take place (the attraction being educational rather than reinforcing the visual consumption of poverty) and the way the tours are organised (which enables the empowerment of tour guides) demonstrate how Unseen Tours remains outside the usual slum tourism definitions. The power to deliver social change has been at the heart of Unseen Tours' work from the very start and the analysis offered in this chapter demonstrates that Unseen Tours can help to effect societal change in the city. Homeless tour guiding needs to be seen alongside other forms of social tourism in cities – which is an under-researched phenomenon. While a range of tourism initiatives with a social purpose do exist in London, little attention has been paid to the role these can play in societal change in the city. The Good Hotel London, for example, trains unemployed locals and stimulates local businesses by being strongly rooted in the local community (Good Hotel London 2018). Their social business concept thus shows strong parallels with Unseen Tours, going beyond the CSR agenda towards more sustainable social change in an urban context. Research to date has, however, not paid much attention to this in the context of London's tourism landscape.

Homeless tour guiding can be a way to create new livelihoods for marginalised members of society, thus making urban tourism more socially inclusive. In this chapter we have argued that homeless tour guiding can give residents political power and create an awareness of a marginalised group of British society. Care should be taken to make sure that social enterprises do not replace government responsibilities and actions. However, organisations like Unseen Tours do constitute a welcome addition to state support and the welfare system and can create more awareness of problems that are often overlooked in the traditional tourism narrative.

By bringing homeless tour guiding into the context of slum tourism and empowerment, this chapter offers one of the first contributions on this phenomenon. Though more research is needed on tourists' motivation to participate in these tours, the tour guides' narratives, how such tours can shape and change people's realities and challenge stereotypes, as well as the role of homeless tour guiding in urban regeneration, this chapter outlines some of the issues missing from existing studies. By understanding homeless tour guiding as a social and a spatial phenomenon which is commercialised for the tourist but designed to help the guide, we will be better able to understand social inequalities in the city and how these might be addressed.

References

Allen, Lissette. 2016. 'Poverty Tourism: Homeless Guides show Prague's less Salubrious Side'. *The Guardian*, 1 June. Available from https://www.theguardian.com/sustainable-business/2016/jun/01/prague-homeless-tourist-guide-pragulic-city-less-salubrious-side

Baptista, João. 2010. 'The Tourists of Developmentourism – Representations "from below"'. *Current Issues in Tourism*, 14(7), 651–667.

Bauman, Zygmunt. 2000. *Liquid Modernity*. Cambridge: Polity Press.

Bland, Archie. 2015. 'The Streets are Ours: A London Walking Tour through the Eyes of the Homeless'. *The Guardian*, 14 March. Available from https://www.

theguardian.com/lifeandstyle/2015/mar/14/london-different-perspective-walking-tour-homeless-guide.

Bornstein, David and Susan Davis. 2010. *Social Entrepreneurship: What Everyone Needs to Know.* Oxford: Oxford University Press.

Burgold, Julia and Manfred Rolfes. 2013. 'Of Voyeuristic Safari Tours and Responsible Tourism with Educational Value: Observing Moral Communication in Slum and Township Tourism in Cape Town and Mumbai'. *Die Erde,* 144(2), 161–174.

Butcher, Jim. 2003. *The Moralisation of Tourism: Sun, Sand… and Saving the World?* London: Routledge.

Butler, Patrick. 2018. 'Rough Sleeper Numbers in England Rise for Seventh Year Running'. *The Guardian.* Available at: https://www.theguardian.com/society/2018/jan/25/rough-sleeper-numbers-in-england-rise-for-seventh-year-running (accessed 19 October 2018).

Bryon, Jeroen. 2012. 'Tour Guides as Storytellers – From Selling to Sharing'. *Scandinavian Journal of Hospitality and Tourism,* 12(1), 27–43.

Cheong, So-Min and Marc Miller. 2000. 'Power and Tourism: A Foucauldian Observation'. *Annals of Tourism Research,* 27(2), 371–390.

Crisis. 2018. *The Homelessness Monitor: England 2018.* London: Crisis.

Day, Jonathon and Makarand Mody. 2017. 'Social Entrepreneurship Typologies and Tourism: Conceptual Frameworks'. In Pauline Sheldon and Roberto Daniele, eds, *Social Entrepreneurship and Tourism: Philosophy and Practise.* Cham: Springer International Publishing.

DeVerteuil, Geoff. 2014. 'Being Poor in the City'. In Ronan Paddison and Eugene McCann, eds, *Cities and Social Change: Encounters with Contemporary Urbanism.* London: Sage.

Dolezal, Claudia. 2015a. *Questioning Empowerment in Community-based Tourism in Rural Bali.* Doctoral dissertation. University of Brighton.

Dolezal, Claudia. 2015b. 'The Tourism Encounter in Community-based Tourism in Northern Thailand: Empty Meeting Ground or Space for Change?'. *ASEAS – Austrian Journal of South-East Asian Studies,* 8(2), 165–186.

Dolezal, Claudia. 2011. 'Community-based Tourism in Thailand: (Dis)illusions of Authenticity and the Necessity for Dynamic Concepts of Culture and Power'. *ASEAS - Austrian Journal of South-East Asian Studies,* 4(1), 129–138.

Dredge, Dianne. 2017. 'Institutional and Policy Support for Tourism and Social Entrepreneurship'. In Pauline Sheldon and Roberto Daniele, eds, *Social Entrepreneurship and Tourism: Philosophy and Practise.* Cham: Springer.

Dürr, Eveline and Rivke Jaffe. 2012. 'Theorizing Slum Tourism: Performing, Negotiating and Transforming Inequality'. *European Review of Latin American and Caribbean Studies,* 93, 113–123.

Dyson, Peter. 2012. 'Slum Tourism: Representing and Interpreting 'Reality' in Dharavi, Mumbai'. *Tourism Geographies,* 14(2), 254–274.

Evolve. 2018. Through my Eyes: Perceptions of Homelessness Available at https://www.evolvehousing.org.uk/campaign/perceptions-of-homelessness/ (accessed 19 October 2018).

Freire-Medeiros, Bianca. 2013. *Touring Poverty.* London: Routledge.

Freire-Medeiros, Bianca. 2012. 'Favela Tourism. Listening to Local Voices'. In Fabian Frenzel, Ko Koens and Malte Steinbrink, eds, *Slum Tourism: Poverty, Power and Ethics.* London: Routledge.

Freire-Medeiros, Bianca. 2009. 'The Favela and its Touristic Transits'. *Geoforum,* 40(4), 580–588.

Frenzel, Fabian. 2014. 'Slum Tourism and Urban Regeneration: Touring Inner Johannesburg'. *Urban Forum,* 25, 431–447.

Frenzel, Fabian. 2013. 'Slum Tourism in the Context of the Tourism and Poverty (relief) Debate'. *Die Erde: Journal of the Geographical Society of Berlin,* 144(2), 117–128.

Frenzel, Fabian. 2012. 'Beyond "Othering": The Political Roots of Slum Tourism'. In Fabian Frenzel, Ko Koens and Malte Steinbrink, eds, *Slum Tourism: Poverty, Power and Ethics.* London: Routledge.

Frenzel, Fabian and Stephanie Blakeman. 2015. 'Making Slums into Attractions: The Role of Tour Guiding in the Slum Tourism Development in Kibera and Dharavi'. *Tourism Review International,* 19(1–2), 87–100.

Frenzel, Fabian and Ko Koens. 2012. 'Slum Tourism: Developments in a Young Field of Interdisciplinary Tourism Research'. *Tourism Geographies,* 14(2), 195–212.

Good Hotel London. 2018. *Mission.* Available from http://www.good hotellondon.com/mission/ (accessed 29 April 2018).

Goodwin, Harold. 2011. *Taking Responsibility for Tourism.* Oxford: Goodfellow Publishers.

Hallin, Anette and Peter Dobers. 2012. 'Representation of Space: Uncovering the Political Dimension of Guided Tours in Stockholm'. *Scandinavian Journal of Hospitality and Tourism,* 12(1), 8–26.

Homeless Link. 2018. Rough Sleeping – Explore the Data. Available at https://www.homeless.org.uk/facts/homelessness-in-numbers/rough-sleeping/rough-sleeping-explore-data (accessed 19 October 2018).

Jarosz, Andy. 2012. A tour of London street survival. BBC. Available at http://www.bbc.com/travel/story/20120627-a-tour-of-london-street-survival.

Jonasson, Mikael and Nicolai Scherle. 2012. 'Performing Co-produced Guided Tours'. *Scandinavian Journal of Hospitality and Tourism,* 12(1), 55–73.

Kassam, Ashifa. 2013. *Poorism, the new tourism: Travellers help the homeless by signing up for tours of the rougher side of town.* Available from http://www.macleans.ca/society/life/poorism-the-new-tourism./

Kinder, Barbara. 2016. 'Devolution of the Welfare State? A Three Country Comparative Analysis of Social Enterprises within Welfare State Models'. *Blue Gum,* 3, 102–118.

Koven, Seth. 2006. *Slumming: Sexual and Social Politics in Victorian London.* Princeton, NJ: Princeton University Press.

Kraftova, Ivana, Zdrazil, Pavel, and Martin Mastalka. 2015. 'The Role of Social Enterprise in Regional Development: The Case of the Czech Republic. *Japan Social Innovation Journal,* 5(1), 14–31.

Lum, Claryce. 2016. '"Arm, aber sexy" (Poor, but sexy): Homelessness in Berlin'. *Ethnoscripts,* 18(1), 42–59.

Mottiar, Ziene. 2016. 'Exploring the Motivations of Tourism Social Entrepreneurs: The Role of a National Tourism Policy as a Motivator for Social Entrepreneurial Activity in Ireland'. *International Journal of Contemporary Hospitality Management,* 28(6), 1137–1154.

McCabe, Scott, Minnaert, Lynn, and Anya Diekmann, eds. 2012. *Social Tourism in Europe: Theory and Practice.* Bristol: Channel View Publications.

MacCannell, Dean. 1999. *The Tourist: A New Theory of the Leisure Class.* Berkeley: University of California Press.

Mekawy, Moustafa. 2012. 'Responsible Slum Tourism: Egyptian Experience'. *Annals of Tourism Research,* 39(4), 2092–2113.

Meschkank, Julia. 2011. 'Investigations into Slum Tourism in Mumbai: Poverty Tourism and the Tensions between Different Constructions of Reality'. *GeoJournal,* 76, 47–62.

Minnaert, Lynn. 2014. 'Social Tourism Participation: The Role of Tourism Inexperience and Uncertainty'. *Tourism Management,* 40, 282–289.

Minnaert, Lynn, Maitland, Robert, and Graham Miller. 2006. 'Social Tourism and its Ethical Foundations'. *Tourism, Culture and Communication,* 7(12), 7–17.

Monroe, Elizabeth and Peter Bishop. 2016. *Slum Tourism: Helping to Fight Poverty…or Voyeuristic Exploitation? Research Briefing.* London: Tourism Concern.

Morris, Hugh. 2015. 'London's Homeless Tour Guides Experience Booking Surge'. *The Telegraph.* Available from http://www.telegraph.co.uk/travel/destinations/europe/united-kingdom/england/london/articles/Londons-homeless-tour-guides-experience-booking-surge./

Mowforth, Martin and Ian Munt. 2016. *Tourism and Sustainability: Development, Globalisation and New Tourism in the Third World,* Fourth edition. London: Routledge.

Novelli, Marina et al. 2016. 'Travel Philanthropy and Sustainable Development: The Case of the Plymouth–Banjul Challenge'. *Journal of Sustainable Tourism,* 24(6), 824–845.

OBrien, Peter. 2011. 'Business, Management and Poverty Reduction: A Role for Slum Tourism?'. *Journal of Business Diversity,* 11(1), 33–46.

Paddison, Ronan and Eugene McCann. 2014. *Cities and Social Change: Encounters with Contemporary Urbanism.* London: Sage.

Pati, Anita. 2010. 'Homeless People Give an Alternative Guide to London'. *The Guardian,* 25 August. Available from https://www.theguardian.com/society/2010/aug/25/homeless-guides-tour-of-london.

Querstadtein. (2016a). Available at http://querstadtein.org/en/about-us/our-story/ (accessed 3 January 2018).

Querstadtein. (2016b). Available at http://querstadtein.org/en/tours/refugees-show-their-berlin/ (accessed 3 January 2018).

Reisinger, Yvette and Carol Steiner. 2006. 'Reconceptualising Interpretation: The Role of Tour Guides in Authentic Tourism'. *Current Issues in Tourism,* 9(6), 481–498.

Rolfes, Manfred. 2010. 'Poverty Tourism: Theoretical Reflections and Empirical Findings regarding an Extraordinary Form of Tourism'. *GeoJournal,* 75(5), 421–442.

Salazar, Noel. 2005. 'Tourism and Glocalization: "Local" Tour Guiding'. *Annals of Tourism Research,* 32(3), 628–646.

Scheyvens, Regina. 2003. 'Local Involvement in Managing Tourism'. In Singh, Shalini, Dallen, Timothy and Ross Dowling, eds, *Tourism in Destination Communities.* Wallingford: CABI.

Scheyvens, Regina. 2002. *Tourism for Development: Empowering Communities.* Harlow: Pearson Education Ltd.

Selinger, Evan. 2009. 'Ethics and Poverty Tours'. *Philosophy and Public Policy Quarterly,* 29 (1/2), 2–6.

Shades Tours. 2018. Available at https://www.shades-tours.com/ueber-uns/ (accessed 3 January 2018).

Sharis, Moshe and Miri Lerner. 2006. 'Gauging the Success of Social Venture Initiated by Individual Social Entrepreneurs'. *Journal of World Business,* 41(1), 6–20.

Sheldon, Pauline, Pollock, Anna, and Roberto Daniele. 2017. 'Social Entrepreneurship and Tourism: Setting the Stage'. In Pauline Sheldon and Roberto Daniele, eds, *Social Entrepreneurship and Tourism: Philosophy and Practise.* Cham: Springer International Publishing.

Smith, Andrew and Ilaria Pappalepore. (2015). 'Exploring Attitudes to Edgy Urban Destinations: The Case of Deptford, London'. *Journal of Tourism and Cultural Change,* 13(2), 97–114.

Taylor, Jerome. 2011. 'Homeless Tours offer Visitors an Alternative View of Life in London'. *The Independent.* Available at http://www.independent.co.uk/news/uk/home-news/homeless-tours-offer-visitors-an-alternative-view-of-life-in-london-2374273.html.

Telfer, David and Sharpley, Richard. (2008). *Tourism and Development in the Developing World.* London: Routledge.

Unseen Tours. (2016). *Unseen tours: about us.* Available at https://sockmobevents.org.uk/about-us/(accessed 7 January 2017).

UNWTO. (2009). *Tourism and Migration: Exploring the Relationship between Two Global Phenomena.* Madrid: UNWTO.

All internet sites without an access date last accessed on 24 July 2018

CHAPTER 8

The River Thames: London's Riparian Highway

Simon Curtis

Introduction

'*Kingdoms may come, kingdoms may go; whatever the end may be, Old Father Thames keeps rolling along; down to the mighty sea*' (Wallace and O'Hogan 1933).

London's famous river has long been the subject of reverence and worship and the deified figure of Old Father Thames symbolises the spiritual hold which this great river has on the city's culture and people. The Thames articulates the city; it is its artery and lifeblood and its most definitive geographical feature. Crossed by 33 bridges, connecting the north and south banks of London, the river offers a lens into over 2,000 years of human occupation. A voyage along the river is a remarkable experience enjoyed by tens of thousands of visitors and Londoners every day, the most evocative symbol of London's stunning heritage and its post-1990 renaissance, re-establishing

---

How to cite this book chapter:
Curtis, S. 2019. The River Thames: London's Riparian Highway. In: Smith, A. and Graham, A. (eds.) *Destination London: The Expansion of the Visitor Economy.* Pp. 165–182. London: University of Westminster Press. DOI: https://doi.org/10.16997/book35.h. License: CC-BY-NC-ND

Figure 8.1: The View of the River Thames from Switch House, part of Tate Modern (Photo: Tristan Luker).

itself, in the eyes of many global commentators, as the world's greatest city (Pricewaterhouse Coopers 2016).

The River Thames is effectively London's largest open space, despite the presence of the city's numerous Royal Parks. Its sinuous path offers intriguing vistas at each gentle turn. It has featured in some of the great works of Shakespeare and Dickens, the poetry of Wordsworth and Blake and the paintings of Turner, Whistler and Canaletto. More recently, musicians have found inspiration in the river, with The Kinks romanticising the view from Waterloo Bridge: '… as long as I gaze on Waterloo sunset, I am in paradise' (Davies 1967).

The Thames is relatively short for such a famous river, running for just 220 miles from source to mouth (Fathers 2015). It is not even the longest river in England: that particular honour belongs to the River Severn. This chapter concerns itself not with the eastern flowing river which passes through the bucolic landscape of the Cotswolds and Berkshire, but rather with the tidal river reaching from the North Sea and the vast Thames estuary in the east to Teddington Lock in the western extremity of London, some 95 miles upriver.

For centuries, river traffic was dominated by the needs of trade, defence of the realm and ferry crossings, but the waters are now busy with river buses, cruise boats and pleasure craft carrying commuters, sightseers and adrenaline seekers. After several decades of under-use between the 1960s and 1980s, the river is once again becoming a busy highway. This chapter describes the changing role of the river over time, over the last 50 years in particular, and outlines how tourism has become a dominant influence; bringing visitors to the riverside, to the bridges of the Thames and onto the river itself. The chapter outlines the river's pivotal position in London's booming tourist economy and the different ways the river is experienced by visitors.

History and Symbolism

The story of the Thames' impact on the development of tourism in London needs to be set in the context of the river's historical and geographical evolution. Over the centuries, the river has shaped the city but Londoners have also shaped the river.

The early centuries of London's development were dominated by the need to establish and protect the emerging settlement (now largely where the financial district – 'the City' – is located). The Romans built a city wall, a fort (the beginnings of what became the Tower of London) and the first bridge across the river (approximately where London Bridge now stands), which was then considerably wider than the river we know today. The River Thames formed part of the defensive alignment of the city and was the link to the sea and the maritime trade which grew to supply the growing city. Travel and trade by river and sea was much easier than by land for many centuries. The river was thus the key to London's birth and the lifeblood of its development.

As London became established as the seat of royalty for the emerging nation, successive monarchs chose to build their palaces alongside the river, which enabled easy access when travelling but meant they also benefited from the riverside defence structures built to protect the city. Several palaces emerged in the centre of the city near London Bridge (Westminster, Whitehall, the Tower itself) and later outlying palaces were built, flourishing in the Tudor period (the expansion of Windsor and the building of Fulham, Greenwich, Richmond and Hampton Court). Some of these sites have largely disappeared and others have been adapted to new uses, but most survive in wonderful condition and provide the main historical attractions for the increasing number of river cruisers.

Royal pageants were common in the later medieval and Tudor eras, as the river was the perfect stage to display the power and extravagance of the monarchy. The royal court was also keen to pay tribute to the river; Old Father Thames was the water god for the rich and powerful as much as the humble Londoner.

Remarkably, London Bridge remained the only bridge crossing of the river for 1,700 years (Port of London Authority, 2018). There was a succession of structures at this crossing point, most notably the great stone medieval bridge which itself lasted for over 600 years. London Bridge connected the fortified City on the north bank to the notorious medieval settlement of Southwark on the south bank, a den of furtive smuggling, drinking, prostitution and crime. The later medieval period was also the era of the 'watermen', the workers who ferried people and goods across the river in small skiffs.

In addition to defence and royal residences, the main historical role of the river has been to aid London's gradual trading and industrial development from the early medieval period through to the mid-twentieth century, when power stations were still being built on the Thames' banks. The Thames was always a river of commerce, from the ship-chandlers of Wapping to the rope-makers of Limehouse and the myriad boatmen: chalkmen, eelmen, lighter-men and water bailiffs (Ackroyd, 2001). Industrial development has left some remarkable buildings along the riverside, but it also signified a long period when London abused, and indeed turned its back on, its river. The Thames was effectively the city's sewer for centuries and the banks became dominated by factories, furnaces and mills, all dispensing their foul waste and chemical pollutants into the river.

William Blake wrote in the late eighteenth century about the 'dark satanic mills' of the Thames riverside, a poem thought to be based on the Albion Flour Mills in Southwark, long since destroyed (Exploring Southwark, 2018). His poem 'London' echoed the brutality of life in that period, dominated by the smog and stench of the river.

> I wander thro' each charter'd street
> Near where the charter'd Thames does flow
> And mark in every face I meet
> Marks of weakness, marks of woe (Blake, 1794)

50 years later, things had barely improved, as can be noted from Dickens' *David Copperfield*:

> The clash and glare of sundry fiery Works upon the river-side, arose by night to disturb everything except the heavy and unbroken smoke that poured out of their chimneys. Slimy gaps and causeways, winding among old wooden piles, with a sickly substance clinging to the latter, like green hair, and the rags of last year's handbills offering rewards for drowned men fluttering above high-water mark, led down through the ooze and slush to the ebb-tide. (Dickens, 1849)

This was the era of the hellish convict ships anchored in the mud of the Thames estuary; a period that famously ended with 'The Great Stink' in the summer of 1858, when the stench from the river and a serious cholera outbreak prompted the government to commission a vast new sewage infrastructure for the city, achieved some 15 years later and overseen by the great Victorian engineer Joseph Bazalgette.

Bazalgette's transformation of London's sewage system involved substantial bank extension and infill, reducing the width of the river in the central part of the city. Gradually, industrial development and dock activity moved to the eastern 'marshes' of the Thames. This was the era of the 'taming' of the marshes and of the Thames' numerous feeding tributaries and inlets. Many tributaries (such as the Fleet, the Tyburn and the Walbrook) were culverted and covered and large docks were built in Rotherhithe, Blackwall and on the Isle of Dogs and beyond. These areas became the new industrial heartland for a century, while the river flowing through Westminster gradually returned to a more serene quality – though it remained a place of commerce rather than recreation in this period.

In the late 1950s, despite the improvements to London's sewage system, the Thames was considered to be biologically dead due to the industrial waste and noxious effluents which continued to be dumped in its waters by riverside factories. Over the last 60 years, the de-industrialisation of inner London, together with legislation and pro-active river management, has meant that the Thames is now acknowledged as one of the cleanest metropolitan rivers in the world (Erfurt-Cooper 2009).

By the late 1970s London's docklands had effectively become abandoned by commercial shipping, yet they have adapted to less obtrusive commercial, leisure and residential uses in the last 30 years, helped by the foresight of Michael Heseltine who, when Secretary of State for the Environment in the early 1980s, created the London Docklands Development Corporation (Schneer 2005). Shortly after this, the inner London boroughs began to view the tourism sector as a potential economic catalyst and planning policy towards the riverside became more focused on recreation and culture, with co-operative initiatives such as the Cross River Partnership.

Gradually, the river began to move away from its industrial past, and in the twenty-first century, the Thames rediscovered its identity as a place of recreation and relaxation. London's rich history and historical structures have become the main ingredient of a thriving tourist industry and the river is less of a trading route and more of a tourist highway, and a visitor attraction in its own right.

A Highway of Attractions

London is a difficult city to navigate for the first-time visitor. With a fairly flat topography, there are few elevated vistas in central districts and, other than the Royal Parks, relatively few large public open spaces and squares. Other great European cities such as Paris, Barcelona or Rome have long straight avenues, higher ground and connected monuments which provide the visitor with navigational aids. London defies logic at street level and also tends to funnel people below ground on the Underground network. The parks offer respite (see Chapter 10) but it is only the river and its bridges that afford wide vistas, views and a sense of how the city fits together. In fact, the river is often so captivating that those riverside areas which are traffic free (most notably the South Bank) are a magnet for visitors.

A remarkable number of London's major tourist attractions are either on or just beyond the river. The list of the ten most visited UK attractions in 2017, produced by the Association of Leading Visitor Attractions (all of which are in London) reveals that five are located on the city's river banks (indicated in Table 8.1 in italics). Two further riverside attractions (Kew Gardens, Tate Britain) were the 14th and 15th most visited in the UK in 2017.

Attraction	Visitor Numbers (2017)
British Museum	5.91m
Tate Modern	*5.66m*
National Gallery	5.23m
Natural History Museum	4.43m
V & A Museum	3.79m
Science Museum	3.25m
Southbank Centre	*3.23m*
Somerset House	*3.22m*
Tower of London	*2.84m*
Royal Museums Greenwich	*2.61m*

Table 8.1: The UK's Ten Leading Visitor Attractions. Source: ALVA, 2018.

Figure 8.2: The River Thames Foreshore looking North towards St Paul's Cathedral (Photo: Tristan Luker).

London's riverside is strewn with major visitor draws from the west (Hampton Court Palace, Richmond riverside) to the east (Historic Greenwich, O2 Arena) with a particularly high concentration of visitor sights between Vauxhall Bridge and Tower Bridge. Palaces, great churches and monumental buildings jostle for superiority here – the Palaces of Westminster, Whitehall, Lambeth and the Tower, the iconic Tower Bridge, a great Abbey, two Cathedrals (St Paul's and

Southwark), a world class arts complex along the South Bank and the cathedral-like Tate Modern. Squeezed in between are the more recent additions to London's tourist scene – the needle-like Shard and the revolving London Eye (see Chapter 6), an aquarium, the London Dungeons, a reconstructed Elizabethan theatre (the Globe) and a Second World War museum ship (HMS *Belfast*).

The Port of London Authority estimated that over £2 billion of GDP was generated by tourism in 2015 in the wards immediately adjacent to the river (Port of London Authority 2016) and that there were 23.4 million visitor trips to attractions beside the Thames in 2015, of which 4.7 million had a direct maritime connection. This reaffirms the significance of the River Thames and riparian territory in London's visitor economy.

London's four World Heritage Sites (the Royal Botanic Gardens in Kew, the Palace of Westminster and the Abbey, the Tower of London and Maritime Greenwich) all flank the riverside and are symbolic of the role the river played in establishing London and Britain's maritime, scientific and global supremacy during the Empire period. As the Mayor of London wrote in 2013:

> The river's vital role as both an artery for transporting people through the heart of London and as a playground for people to explore the wonders of the city are on show for the world to see. A trip along the Thames reveals 2,000 years of riparian history; from the Roman walls at Tower Hill, and the Victorian wharves and warehouses to the soaring peak of The Shard – providing a stunning vista of London's past and present. (Mayor of London, 2013)

In the twenty-first century we have started to see the development of further outlying tourist attractions along the eastern banks of the Thames in London, and indeed its prime remaining tributary (the Lea), as regeneration of the docklands area enters a mature phase. The O2 Arena, the recently restored and re-presented *Cutty Sark*, the Museum of London Docklands, the Crystal, the Thames Barrier and The Orbit at the Queen Elizabeth Olympic Park are all relative newcomers to the exalted highway of riverside attractions in London.

The River as Tourist Space and Visitor Experience

The Thames has been a recreational resource for centuries, even if this role was undermined at the height of the city's industrialisation. With the global development of urban tourism in recent decades, and by virtue of the plethora of tourist attractions and sights, the river has become a thriving tourist space. This is immediately evident when standing on any of the central London bridges and surveying the river craft which pass by. Boats and vessels now predominantly carry sightseers and river piers are cluttered with tourist infrastructure. A walk across these bridges is inevitably interrupted by visitors posing for

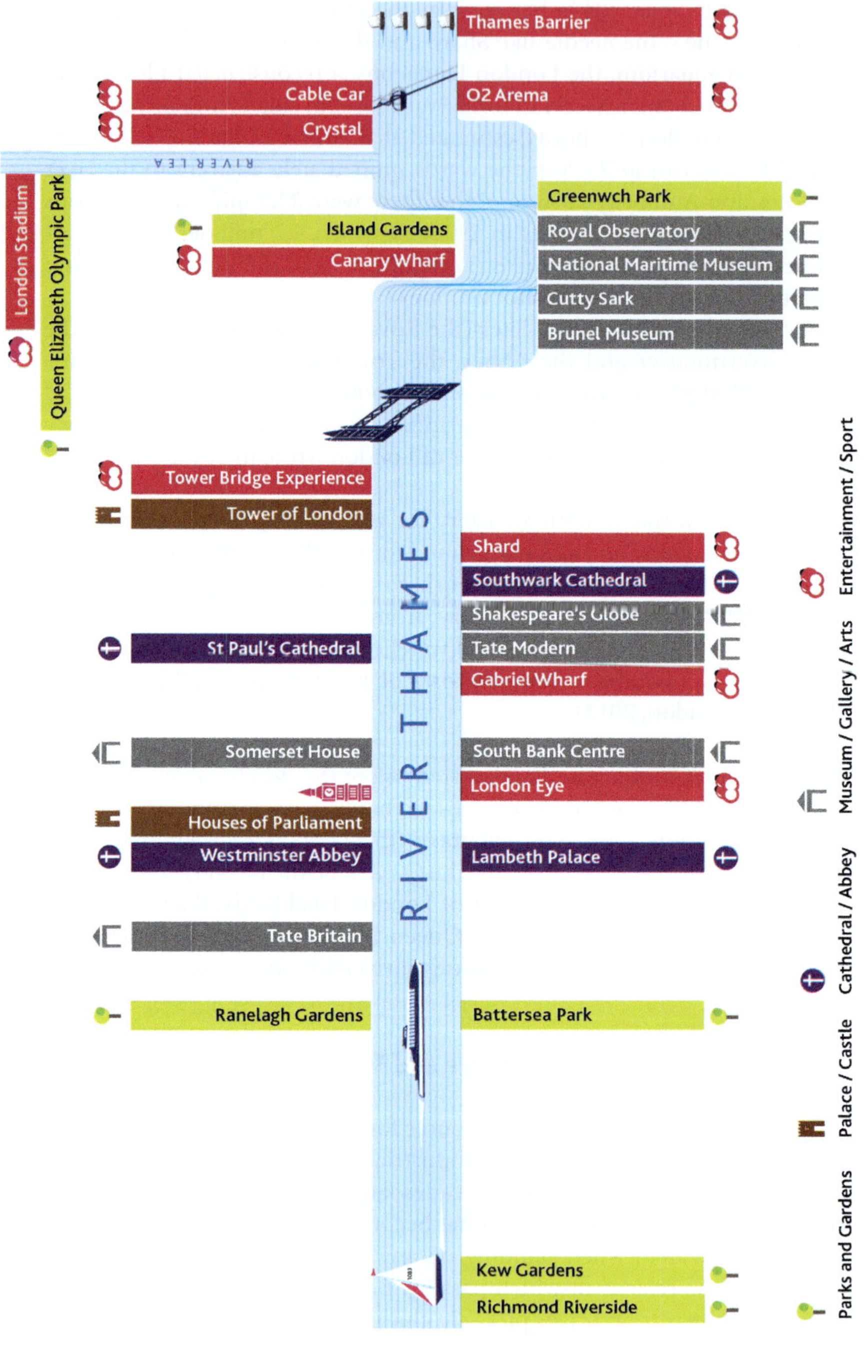

Figure 8.3: A Graphic Illustration of London's Tourist Highway (©Mason Edwards).

photographs. Pedestrianised riverside promenades (especially along the South Bank) are almost permanently inhabited by buskers, performers and pop-up retailers.

City tourism has thrived globally over the last 30 years, driven by a number of supply factors such as civic cultural investment, new and exciting accommodation options, low cost airlines, new events and festivals and by new digital communications. Demand factors have arguably been even more influential as cities have become the key beneficiary of the popular fashion for taking short breaks and discovering new cultural experiences through visiting different cities.

The urban tourism experience can be analysed from a number of alternative perspectives. Two ideas in particular, both heavily embedded in geographical and tourism academic thinking in recent decades, offer useful frameworks through which to consider the role of the Thames in the London visitor experience. John Urry explored the tourist experience through the idea of the tourist gaze (1990; 2002) whereby the visitor is ultimately in search of visual messages, symbols and memories, especially those than can be easily photographed and represented. MacCannell (2001) took Urry's concept further and noted that some places and some situations could inspire a 'second gaze', one that is more satisfying, deeper and reflective. There are few locations in London that can engender the 'second gaze' but the Thames riverside is one of them. It is where the visitor can make sense of London and where great sites can be seen as part of a wider cityscape. On occasions, beneficial weather and light conditions may see the cityscape reflected and refracted in the river water. The meandering nature of the river through central London can lead to abstractions and disorientation in the appreciation of the cityscape, adding more depth to the 'second gaze'. This links to the second conceptualisation of the visitor experience: the occasions when an experience may become fully 'embodied' and sensory. Fully embodied experiences may be fleeting, but they are rich, multi-dimensional and shape memories and attachments (Crouch 2000).

The Thames and its riverside host alternative experiences that suit the visitor's moods and desires, and have the ability to inspire a fully embodied visitor reaction. Waterways offer a means of transport between sights and this is often the most visceral method for moving through the cityscape. They also offer opportunities for adventure through the many adrenaline fuelled rides that are now available on the water. The river can be a place for reflection and contemplation, offering gentle riverside walks and traffic-free spaces to take in views; and it provides numerous attractive places for eating, drinking and socialising; ultimately, it provides diverse views and widescreen vistas and a pleasing setting for the numerous cultural and historic sights along its course. It is worth considering each of these experiential dimensions in more detail.

Transport for London (TfL) estimated that 10.3 million passenger movements took place on the river in 2015/16 (Transport for London, 2018) and have projected an increase to 12 million movements by 2020. These figures include river bus services and the Woolwich ferry and do reflect a gradual increase in tourist boat and cruise traffic over the last decade. The river bus service, run by MBNA Thames Clippers and licensed by TfL, now comprises 17 catamarans and is responsible for 4 million passenger movements (Thames Clippers, 2018), a remarkable statistic for a service only launched in 1999. Though ostensibly a commuter service, the company has embraced sightseers and, increasingly aware of the appeal of its service to tourists, has introduced a visitor audio guide onto its boats.

Other boat services are more dedicated to visitors with companies such as City Cruises and Thames River Services offering a range of short hop-on/hop-off circular cruises as well as longer cruises to Greenwich and the Thames Barrier. More leisurely dinner cruises, event cruises and charter services are offered by all of the main boat operators. There are even cruises where passengers can watch movie screenings. TfL list over 40 operators for private hire and charter on their web site (Transport for London, 2018), and there is evidence of consistent growth in the popularity of corporate hire and private party hire. Over the last decade, the traditional cruising boats have been joined by several operators of speed boats and RIB boats along the river, thrill rides offering high speed twists and an adrenaline oriented, fully embodied experience.

The Thames can be walked from source to the Thames Barrier, a 184-mile route which was given national trail status in 1989 (Fathers 2015). The London Thames Path, downriver from Teddington, splits into two paths either side of the river. The paths on the north and south shores through the capital occasionally veer away from the riverside and at times abut very busy roads, but it is largely a consistent and rather magnificent urban trail. For many, this is the way to enjoy the river; walking several miles along the path enjoying surprising views of familiar and unfamiliar buildings and observing the recent transformation of stretches of formerly industrialised riverside. In places, the walk can feel more like a promenade as the path veers through Battersea Park and along the South Bank into Bankside, the most popular and animated section. The practice of promenading is perhaps returning to London, something which had its heyday in the eighteenth and nineteenth century pleasure gardens of Chelsea (Cremorne and Ranelagh) and Vauxhall (Fathers 2015).

Perhaps some of the most satisfying strolls along the Thames Path are to be had to the west of the city, in Isleworth, Richmond, Kew, Chiswick and Hammersmith, where some of London's best pubs have flourished due to their unique and coveted riverside settings. Historically, many breweries were located alongside the river as ready access to a water source was a necessity and pubs tended to congregate in close proximity. Many a story of the river's ghoulish and dark past is proudly told or displayed in such iconic riverside pubs as the Prospect of Whitby in Wapping, the Trafalgar Tavern in Greenwich, the Dove in Hammersmith and the London Apprentice in Isleworth. In more recent times, restaurants and café operators have sought to take advantage of riverside settings, a boon for business in the summer months.

Much of the success of riverside dining and drinking is related to the sense of nature which the river brings to contemporary urban dwellers and weary visitors. While the city changes and buildings are transformed or redeveloped, the river (though tamed) is an enduring link to the countryside beyond and the place which preceded the claustrophobia of twenty-first century urbanism. It is the views and the setting which the river provides which have fuelled its pivotal role in London's touristic expansion. The panoramic tourist gaze (see also Chapter 6) is facilitated by the numerous bridges but also increasingly by public viewing galleries in towers and visitor attractions; and in Docklands, by the new cable car 'airline' which flies between Greenwich Peninsula and the Royal Docks.

The River as Stage

The Thames in London has long been the city's primary public stage. It has served as the venue for royal showmanship, for pageantry and carnival, for national celebrations, for sports and racing, and for cultural expression. This role has been consistently exploited by the city's tourist sector in recent times and extends to the river banks with open-air events and street theatre offered at the South Bank Centre, City Hall, the Tower of London and the Royal Hospital in Chelsea (host of the annual Chelsea Flower Show). More recently, London has adopted the European trend of creating a riverside beach in the summer months.

Since the Tudor period, the river has often been chosen as a route for royal celebration and to mark coronations and anniversaries, personified by the reigning monarch of the time travelling in front of a flotilla in the royal barge (Port of London Authority, 2018). For over 600 years, the newly elected Lord Mayor of London travelled by river from the City to Westminster to pledge loyalty to the monarch, as requested by King John in 1215 in the original royal charter granting freedoms to the city barons. Though the procession now takes place on the streets, the Lord Mayor is still accompanied by liveried watermen (Ackroyd 2007). These traditions continue to this day, with the present Queen having celebrated both her Silver and Diamond Jubilees with a river pageant. The latter took place in June 2012 with a partial recreation of Canaletto's famous Thames paintings as 670 military, commercial and pleasure craft processed downriver from Wandsworth to Tower Bridge attracting an estimated one million spectators (Port of London Authority, 2018). The Queen's Royal Household still appoints a Royal Bargemaster and 24 Royal Watermen (Royal Household, 2018), a link to the past importance of the river for royal transport but also an important symbol of the river's continuing use for ceremonial events.

For a period between the early seventeenth and early nineteenth centuries, the river also played an occasional and unlikely role as host of the infamous Frost Fairs (see Chapter 9). This was the period of the 'Little Ice Age' when winter temperatures were much lower than today and the Thames froze over for up to two months at a time (Historic UK, 2018). The freezing of the river tended to happen just upstream from London Bridge (the only bridge until 1726), as the bridge acted to slow the flow of the river, which was much shallower before Bazalgette's infilling and narrowing. The temporary Frost Fairs became increasingly popular and carnivalesque with an almost bacchanalian spirit which included skating, puppet plays, food stalls, bear-baiting and general drunken revelry (Fathers 2015). The river acted as a liminal space allowing the populace to create festivity and spontaneous theatre.

In more recent times, the river has become the focus for London's renowned New Year's Eve celebrations, rivalling those in Sydney, New York and Paris. Since the early 2000s, the Victoria Embankment and the South Bank area of the riverside have become the focus for fireworks displays to mark the start of the New Year, utilising the bridges and reflections provided by the river. The popularity of the event has meant that the best viewing areas are now ticketed, providing a further example of the commercialisation of London's public space noted in Chapters 9 and 10.

Perhaps the most notable new celebration of the river is the Totally Thames Festival, a month-long festival held each September which aims to highlight the diversity and emerging creativity of the riverfront and promote the role of the river in London's cultural fabric through installation art, projections, music and talks (Totally Thames Festival, 2018). The Festival began in 2014, emerging from an earlier cultural event (the Thames Festival). Now organised by the Thames Festival Trust, over 150 separate events are held during the festival month.

As a sporting stage, the Thames again boasts a long and proud history. The noble tradition of the watermen was influential in the recreational popularity of rowing which was established enough by the eighteenth century for the first organised regattas to be held, followed by the formation of numerous rowing clubs, especially in the relatively calmer waters of West London. The historic Doggett's Coach and Badge Race claims to be the oldest rowing race in the world (Port of London Authority, 2018), first contested in 1715, and staged every year since, other than its suspension during World War Two. The event still follows its original course, from London Bridge to Chelsea, and involves a straight race between six apprentice watermen.

The most famous race is of course the annual Boat Race between the 'dark blue' and 'light blue' crews of Oxford and Cambridge Universities respectively, first held on the Thames in 1836, which takes place in late March or early April, between Putney and Mortlake. The women's race, not established until the mid-1960s, now takes place on the same day. The Race attracts in excess of 250,000 spectators to the riverbank and now includes several big screens and a 'Fan Park' but it has long been a national institution as a mass television event, watched by up to 10 million people (Boat Race, 2018). As such, it is treasured not so much for the race but for the occasion and for its traditions and familiarity in the British public consciousness.

The other long-established rowing race is The Heads of the River Race (since 1926) which is rowed in March over the same championship course as the Boat

Race, but in reverse. This is a semi-professional event involving the best rowers from around the world, in crews of eight. In more recent times, another race, the Great River Race, has become a popular spectator event, a river marathon held in September, inviting boats of all shapes and sizes to race 26 miles upriver from Millwall to Ham. Now attracting over 300 boats, this is a colourful mass participation event which has been embraced by Londoners and shows that the river can be enjoyed by enthusiasts as well as elite sportsmen.

The Thames now plays host to more than 80 major sporting events each year (Port of London Authority 2016). The re-invention of London's docks as recreational spaces is enabling additional water-sports to be offered, such as water-skiing, paddle boarding and wind surfing. There are now two specialist water-sports centres, at Millwall Dock and at Royal Victoria Dock. The latter hosted Formula One powerboat racing in June 2018, emphasising the Thames' ability to adapt to new and more specialist participation and spectator sports.

The river as stage is perhaps most intensively experienced at the South Bank complex. This array of cultural buildings set in a pedestrianised environment overlooking the river is a legacy of the Festival of Britain in 1951. The spirit of celebration and showmanship of that event has continued over the years and has indeed recently spread eastwards to Bankside, helped by the establishment of cultural attractions such as Gabriel's Wharf, Tate Modern, and the Globe Theatre. This is now a destination of choice for Londoners and visitors who want to enjoy the riverside in a traffic free environment with rather unexpected and spontaneous 'fringe style' entertainment and cultural expression. The South Bank has indeed transformed the established tourist walking routes across central London in recent years. It has become a hugely successful place, especially considering that, as recently as the mid-1990s, Southwark Council was bemoaning the lack of riverside eating and drinking opportunities along the South Bank. Its evolution has involved extensive collaboration between a complex network of local partnerships. A combination of well-designed public realm, some fortuitous neighbouring cultural investment and determined involvement of local stakeholders have contributed to this success (Tyler and Morad 2008).

The most recent addition to the South Bank's allure is its regular summer programme of outdoor events based around the creation of a riverside beach scene. Taking inspiration from the success of the 'Paris Plages' along the Seine and from urban beach creations in cities such as Berlin, the South Bank initiative is now extending to other parts of the riverside, including the Royal Docks and Fulham.

The river has been chosen as a setting for countless classic scenes in film and television in recent decades. This has added to its iconography and has been a telling factor in attracting more international tourism to London. The river has staged thrilling chase sequences in recent Bond films and, further back in time, it featured memorably in the Beatles' *A Hard Day's Night*, in *A Clockwork Orange* and in a touching romantic scene, filmed on the South Bank, in *Four Weddings and a Funeral*. It also featured in two supremely executed scenes in the Harry Potter film series; it was flown over by Harry Potter and friends in an immersive flying sequence *from Order of the Phoenix* and was then shown in a sinister light as the Millennium Bridge was destroyed by death eaters in the opening scene of *The Half-Blood Prince*.

Contested Development

One of the consequences of rapid urban tourism growth is uncontrolled commercialisation and low-quality clutter around the fringes of the spaces colonised by visitors. Writing in 2004, Shaw and Williams reflected on the way in which tourism tends to be driven by globalised trends of consumption and exploitation, and that commodification of place can become ingrained without political will and regulation. Other commentators have further questioned the trend in some cities for the urban tourism experience to be increasingly manipulated and 'experientialised' to maximise consumptive behaviour (Smith 2015). Such commodification has arrived along parts of the Thames in London, most specifically around the most well-used river piers and the London Eye. Here perhaps there are signs of touristification at the cost of the authentic riverside visitor experience and the beginnings of a tourist enclave (Judd 2003). However, few observers would consider it to be an issue of major concern, at least not yet. It imposes no more or less touristification for instance than the West End theatre district or the museum district of South Kensington. London

Figure 8.4: More London – The South Bank between London Bridge and Tower Bridge (Photo: Eman Mustafa).

has an uncanny ability to absorb tourist pressures, in a way that seems to elude smaller capitals like Berlin and Lisbon.

The boom in London real estate and marketability of riverside living has been more controversial. Significant stretches of the river have become colonised by exclusive and expensive apartment blocks, restricting opportunities for re-introducing trees, planting and recreational areas. Though these can present some intriguing and glamorous sights for the river cruisers, they make the river inaccessible in some neighbourhoods, along both banks.

There are improvements to be made in terms of the river's effectiveness as a tourist space; motorised traffic still dominates too much of the north bank, and there are insufficient pedestrian-only bridges. There remain some frustrating stretches of the Thames path where access to the riverside has been prevented either by legal obfuscation, gated new development or by historic warehouses which hug the river-shore.

Despite these issues, the gradual imposition of tourism and the evolution of the river and riverside into a space, or series of spaces, dominated by recreation and tourism, has generally been a positive and enriching process. The Thames has thrived and become perhaps the most important feature of London's overall visitor package; a place of wonder, refuge and escapism for tourists and Londoners.

The Future Role of the River in Destination London

The River Thames features prominently in the numerous strategies and plans published by the Mayor, the Greater London Authority, the Port of London Authority and London and Partners, the Mayor's official promotional agency. The key policies common to these plans are: making the riverside more accessible; to develop further recreational infrastructure; enhancing biodiversity of the river; and boosting the role of the river as a means of transport for residents and visitors. This last policy will involve a commitment from all the leading stakeholders to better integrate river transport with underground and road systems. London and Partners' Tourism Vision for London (2017) does not make specific recommendations in relation to the river but references its vital role in the future development of new visitor experiences and outlines the essential goal of better integrating river and land-based transport.

The Port of London Authority released a comprehensive vision for the river in 2016 called Thames 2035. They project continued growth for port activity and river freight journeys along the river, but most of the vision concerns aspirations to improve pier infrastructure, increase moorings, and develop further sporting opportunities; continued enhancement of the river ecosystem and enhancing appreciation of the river's cultural and educational resources. A key target is to double the number of people travelling each year along the river to 20 million by 2035.

There have been some innovative new projects for river development which have failed in recent years. A project to establish an extensive floating boardwalk with event pavilions and a lido on the north bank between the Millennium Bridge and Tower Bridge was mooted in 2012 but did not progress. One project touted as a way to deliver a number of strategic aspirations was the Garden Bridge scheme, a proposal to build a pedestrian only bridge across the river between the South Bank and Temple, a structure designed by Thomas Heatherwick with more than 270 trees and 100,000 plants linked by meandering paths across a wide bridge. The Bridge was set to become another signature attraction for London, combining infrastructure and an experience celebrating the river's transformation into an environmentally clean and biodiverse resource. The idea was that the garden spaces would change as the pedestrian walked over the bridge, reflecting the character and rich heritage of the riverbanks and offering different experiences as the planting and textures change during the seasons. The project had many high-profile supporters, but was killed off by the election of Mayor Sadiq Khan who was unprepared to underwrite the costs. Investigations are ongoing to work out what happened to the tens of millions spent on consultancy fees for a project that was never built. However, it is unlikely that we have heard the last of new projects to bridge the Thames. The river has a habit of inspiring ambitious projects.

Other new projects are emerging. Battersea Power Station, a cathedral of the industrial age, is being re-invented as a twenty-first century mini-metropolis and there are bold proposals to create a floating village offering ice skating and a lido on Royal Victoria Dock. Also in the Royal Docks, further development of new commercial facilities is likely to see Docklands cement its position as a hub for business tourism, building on the success of City Airport and the ExCel exhibition and conference centre. Meanwhile, the Rothschild Foundation is funding much of a £20 million scheme to permanently illuminate London's 15 central bridges based on designs by US artist Leo Villareal (Rothschild Foundation, 2018).

The prospects for the river's wildlife are perhaps one of the most important aspects of the river's future tourism potential. There are already an estimated 125 types of fish in the tidal Thames (London Wildlife Trust, 2018) and seals, dolphins and porpoises are increasingly spotted swimming upriver towards the city. Investment in wetland nature reserves adjacent to the river bank such as those in Barnes (the London Wetland Centre), Greenwich and Crossness, are encouraging further diversity in the river's bird population. The completion of the Thames Tideway Tunnel in 2021, which will reduce the sewage discharge into the river and its tributaries to virtually nil, will further enhance the cleanliness of the river. Nature and tourism will need to find a way to co-exist in harmony for mutual benefit.

Conclusions

The river rolls in from the sea and up to London with the tide; at the other end it rises in a field near Cirencester to dampen the green grass in a dark curving line that soon becomes a stream. The salt water from the sea and the clear sweet water from the low Cotswold hills meet at Teddington (from 'Conclusion': Schneer 2005).

London is often described as a city in constant flux, but some things have not changed, and the geography of the river remains its defining characteristic. The meeting of the tidal Thames and the sweet water from the bucolic landscape of middle England has been the symbolic heart and the lifeblood of one of the world's great cities. London's riparian spaces have provided secure and defendable refuges, routes for trade, the fuel for industry and a recreational setting. In the last few decades, the Thames and its riverside have adapted to a new era, one which involves a complex array of roles, with tourism bringing much of the animation, verve and spirit to its new persona. The Thames is the artery which winds a sinuous path through an urbanised landscape, connects historic monuments and cultural landmarks in a sensory way, provides breathing space for visitors and a space in which creative expression and festivalisation can prosper. It might be recognised as an exalted highway, allowing residents to become tourists in their own city and offering new perspectives to the returning visitor, providing the excitement of travelling 'the open road' through an ever-changing cityscape.

The River Thames is a special river which stirs emotions; it will continue to lift the hearts and minds of many millions of visitors and residents alike in the decades ahead, inspiring a second gaze, perhaps a Waterloo sunset, and an array of satisfying urban tourism experiences.

References

Ackroyd, Peter. 2001. *London: The Biography*. London: Vintage Books.

Ackroyd, Peter. 2007. *Thames – Sacred River*. London: Vintage Books

ALVA Association of Leading Visitor Attractions. 2018. Available at http://www.alva.org.uk.

Blake, William. 1794. 'London' (part of *Songs of Experience.*)

Boat Race, The. 2018. The Boat Race. Available at http://www.theboatrace.org.

Crouch, David. 2000. 'Places Around Us; Embodied Lay Geographies in Leisure and Tourism'. *Leisure Studies*, 19(2), 63–76.

Davies, Ray. 1967. 'Waterloo Sunset' (song), London: Pye Records.

Dickens, Charles. 1849. *David Copperfield*

Erfurt-Cooper, Patricia. 2009. 'European Waterways as a Source of Leisure and Recreation'. In Prideaux, Bruce and Malcolm Cooper, eds, *River Tourism*, Wallington: CABI.

Exploring Southwark. 2018. Exploring Southwark and Discovering its History Available at http://www.exploringsouthwark.co.uk.

Fathers, David. 2015. *The London Thames Path*. London: Frances Lincoln.

Head of the River Race. 2018. About HoRR. Available at http://www.horr.co.uk.

Historic UK. 2018. Your Guide to the History and Heritage of Great Britain. Available at https://www.historic-uk.com.

Judd, Dennis. 2003. 'Visitors and the Spatial Ecology of the City'. In Lily Hoffman, Susan Fainstein and Dennis Judd, eds, *Cities and Visitors: Regulating People, Markets and City Space*. Oxford: Blackwell.

London and Partners. 2017. *Tourism Vision for London*.

London Wildlife Trust. 2018. Love London. Love Nature. Available at https://www.wildlondon.org.uk

Lord Mayor's Show. Celebrating 800 Years of Paranoia, Murder and Betrayal. 2018. Available at https://www.lordmayorsshow.london.

MacCannell, Dean. 2001. 'Tourist Agency'. *Tourist Studies*, 1(1), 23–37.

Mayor of London. 2013. *River Action Plan*, Transport for London.

Port of London Authority. 2016. *Thames 2035 – Vision for the Tidal Thames*.

Port of London Authority. 2018. About the PLA. Available at http://www.pla.co.uk.

Pricewaterhouse Coopers. 2016. *Empowering City Brands: Bridging the Perception and Reality Divide*. Consumer Intelligence Series.

Rogers, Richard. 2017. 'The Garden Bridge Can Help the City Retain its Global Status'. *London Evening Standard*, 22 June.

Rothschild Foundation. 2018. Winner of Illuminated River International Design Competition announced. Available at https://www.rothschildfoundation.org.uk.

Royal Household, The. 2018. The Home of the Royal Family. Available at https://www.royal.uk.

Schneer, Jonathan. 2005. *The Thames: England's River*. London: Little Brown.

Shaw, Gareth and Allan Williams. 2004. *Tourism and Tourism Spaces*. London: Sage.

Smith, Melanie. 2015. *Issues in Cultural Tourism Studies*, 3rd Edition. London: Routledge.

Thames Clippers. 2018. About MBNA Thames Clippers. Available at https://www.thamesclippers.com.

Totally Thames Festival. 2018. Available at http://www.totallythames.org.

Transport for London. 2018. River. Available at https://www.tfl.gov.uk.

Tyler, Duncan and Munir Morad. 2008. 'London's South Bank Precinct: From Post-industrial Wasteland to Managed Precinct'. In Hayllar Bruce, Tony Griffin and Deborah Edwards, eds, *City Spaces –Tourist Places*. Oxford: Butterworth–Heinemann.

Urry, John. 2002. *The Tourist Gaze*, 2nd Edition. London: Sage

Wallace and O'Hogan. 1933. *Old Father Thames* (song). Chappell Records

All internet sites last accessed on 4 December 2018

CHAPTER 9

Festive Space and Dream Worlds: Christmas in London

Adam Eldridge and Ilaria Pappalepore

Introduction

Mark Connelly, in his social history of Christmas, begins by arguing that Christmas is 'England's single greatest cultural export' (2012, iv). Though aware that modern day Christmas finds its roots across Europe (McKay 2008) and draws on a range of local Pagan and Christian traditions (Miller 1993), Connelly argues that the Victorian period saw a search for and consolidation of uniquely English literature, myths, stories and practices that have since come to define Christmas across the world. Rather than the commonplace argument that the Victorians invented Christmas, Connelly, much like Storey (2008), suggests they were instead concerned that the ritual was dying out and 'conserved' and 'revived' rather than originated an entirely new set of customs and practices. Against the backdrop of rapid social and economic changes such as urbanisation, commercialisation, and industrialisation the Victorians became, for Connelly, 'obsessed' with finding and compiling English customs from the past in order to save this annual festival. This endeavour was especially driven by the desire to unearth and maintain ancient customs unsullied by excessive commercialism and modernity. The emerging materialism associated with

---

How to cite this book chapter:
Eldridge, A. and Pappalepore, I. 2019. Festive Space and Dream Worlds: Christmas in London. In: Smith, A. and Graham, A. (eds.) *Destination London: The Expansion of the Visitor Economy.* Pp. 183–203. London: University of Westminster Press. DOI: https://doi.org/10.16997/book35.i. License: CC-BY-NC-ND

Christmas in the Victorian period, the pantomimes, shop window displays and shopping itself, thus came to be legitimated by the 'aura' and 'antiquity' of Christmas.

Connelly's (2012) account raises two key points for this chapter. Firstly, rather than searching for some fixed, originary truth to Christmas, his argument reminds us that Christmas has long been a compilation of different myths, stories, traditions and rituals. It has also been celebrated, and indeed ignored, in different ways across time. Second, and related to this point, debates about commercialisation, authenticity and the legitimacy of certain practices are not unique to contemporary culture. While bearing these two points in mind, this chapter argues that we have recently seen an intensification and expansion of activity around the festive period, in both spatial and temporal terms. Despite frequent claims that there is a 'war on Christmas', one being fought by those wishing to remove the sacred nature of the festival from the public realm (Feldman 1997; Davis 2010), Christmas is now an integral part of London's entertainment and tourist offer. Christmas Day itself remains a predominantly domestic and family occasion in London, as it does across the UK, but the months leading up to Christmas have now become a period of much cultural activity in the capital. According to data from the 2015 Day Visits Survey, for example, during the month of December, London received 32.7 million day visitors[1], an increase of 18.5 per cent in comparison with the month of November and a 29 per cent increase over January and February. Even more strikingly, overnight domestic visits to London were up by 59 per cent in December 2015 in comparison with November, January and February of the same year[2]. In a report by the New West End Company (NWEC 2016), in the Business Improvement District covering Regent and Oxford Streets, footfall during the festive period was projected to be 30 per cent higher than at other times of the year, while passenger figures to nearby tube stations typically increase by between 20 and 40 per cent. In the six weeks preceding Christmas, West End shops' till receipts were also forecast to rise by £2.34 billion, and employment by some 4,500 to cater for increased footfall and extended opening hours (NWEC 2016).

Other chapters focus on the spatial expansion of tourism; this chapter highlights the way tourism has also extended temporally – into different parts of the year (November–January) and into different times of the day (after dark). The turning on of the Christmas lights on Oxford and Regent Streets in London's shopping heart begins a period that is now characterised by office parties, lighting displays across other shopping streets, Christmas themed pantomimes, films, plays and ballets, Winter Wonderland in Hyde Park, Christmas carols and the lighting of the Christmas tree in Trafalgar Square, Winterville on Clapham Common, Winter Festival on the South Bank and the opening of numerous temporary ice rinks. Theme parks such as Warner Brothers' Harry Potter World have their own Christmas themed displays and then there are the extended shopping hours, Boxing Day sales, mulled wine in many pubs and bars, the media focus on the Christmas number one single, and the fireworks festival on New Year's Eve. Other festivals such as Diwali and Bonfire Night, the more recent addition of Lumiere to the cultural calendar (an illumination festival held in January), and in some cases the celebration of Chinese New Year (Bell 2009), effectively extend Christmas festivities and other cultural events from November to early February. While we can only speculate as to what extent these events, both secular and sacred, from 'here' and elsewhere, represent London's diversity, they do indicate the ways London's Christmas offer mirrors that cobbling-together found in the Victorian period. In turn, we now have a period of several months, all under the auspices of Christmas, that brings together and celebrates a range of different events, traditions, and ways of experiencing the city.

This chapter starts by exploring the history and leverage of Christmas. How this annual ritual has developed and expanded, and the extent to which it has become an important component of London's tourism offer is examined. Its status as a liminal time-space, distinct from the everyday, leads us into a discussion of theming and the wider issue of authenticity and the staging of cultural events for commercial purposes. While, as indicated above, debates about the commercialisation of Christmas are not new, the extent to which these concerns intersect with more recent debates about the privatisation of public space points to an important theme of this chapter (see also Chapter 10). Christmas in London now takes place across a number of private, public, and semi-private spaces. As well as the Christmas lights in central London often being tied in with Hollywood films, large parts of the city's semi-public spaces such as Hyde Park, Leicester Square, Trafalgar Square, and the South Bank become heavily commercial, featuring themed markets, ice-rinks and lighting displays. Many of these events, as explored below, are also commissioned or directly managed by partnerships between local councils and Business Improvement Districts; business-led organisations covering specific geographical areas. After exploring these debates, the discussion turns to more recent work on illumination and the production of Christmas 'atmosphere'. What we find is a very different type of night-time economy developing over this period; similar in terms of the commercial focus and use of alcohol, but quite distinct in terms of being more family-focused and embedded in discourses of pleasure, ritual and tradition. The chapter, in summary, explores the expansion of London's Christmas offer and, echoing Connelly (2012), suggests we are not so much witnessing an entirely new series of behaviours and practices here, but rather an expansion, deepening, and consolidation of both Christmas rituals and debates about the 'proper' meaning of Christmas.

Christmas Spirit and Leveraging

The Christmas experience is multi-sensual and multi-sensory, from the smell of mulled wine to the tactile feeling of unwrapping gifts (Makulski 2015).

Evidence from the UK Office for National Statistics even suggests that Christmas is the most popular time of the year to conceive (BBC News, 2015). The sensuous (Crouch and Desforges, 2003) and embodied (Jokinen and Veijola, 1994) nature of Christmas celebration is therefore key to understanding this annual festival. Hedonistic consumption, including shopping, drinking and eating to excess, plays an important role, often explained by reference to the elusive notion of 'Christmas spirit'. Clarke, in his analysis of the meaning of this concept, defines Christmas spirit as an attitude involving 'a combination of bonhomie, dejected and gay abandon feelings', which is further associated with goodwill, generosity and altruism (2007, 10). In parts of the world where Christmas is widely celebrated, such as Europe and the USA, most businesses and voluntary organisations attempt to leverage the sense of celebration linked to Christmas, and to benefit from association with the 'Christmas spirit'.

The concept of event leveraging refers to those activities that are planned around an event by a sponsor or a host region to maximise potential positive outcomes (Chalip 2004). Chalip's analysis (2006) is very relevant to our discussion, as it links the opportunity for leveraging events with their 'sense of celebration' and 'sense of sharing' or 'camaraderie', two important characteristics of Christmas festivities. According to Chalip (2006), effective strategies to lever the 'emotional and symbolic power' of events (Smith 2014a) include facilitating sociability and informal social opportunities, producing related ancillary events, and theming. As the remainder of this chapter highlights, all these leveraging tools are in evidence during the Christmas period in London.

Pretes (1995) illustrates how the iconic image of Santa Claus was successfully levered to promote Lapland in Finland as the 'real home' of Santa in the 1980s and 1990s. Pretes notes that the Santa Claus village in Lapland offers tourists the opportunity to consume the spirit of Christmas (and associated positive emotions linked to family and childhood) as a cultural commodity. He refers to Santa Claus as a product of the western, postmodern society of spectacle, 'a simulacrum, a copied image for which no original exists' (1995, 14), but a more contemporary analysis would highlight that authenticity is subjective, symbolic and socially constructed. Culture is not static but rather in constant development, which means that objects can become authentic over time (Cohen 1988). For example, Cluley's research (2011) shows how both children and adults suspend their disbelief in Santa in order to enjoy him as if he were real. More importantly, following Wang (1999), authenticity can be seen to be about individual experiences and the state of mind they facilitate, rather than about the actual cultural object and how it is perceived. If visiting Santa's village in Lapland, shopping in a Christmas market or taking a photo with Santa at the shopping mall can help people experience spiritual or aesthetic nourishment (Wang 1999), then that cultural experience is, indeed, authentic. Creative participation is likely to further enhance this type of 'existential authenticity' (Wang 1999): this is perhaps why many people are more likely to engage in creative activities during the Christmas season such as decorating, drawing, crafting, acting or dressing up. Rippin's analysis of Christmas headgear at office Christmas parties (2011), for example, shows the importance of these symbolic head decorations in providing relaxation and a sense of celebration and transgression. Paper crowns, Santa hats, tinsel halos and reindeer antlers all contribute to bringing the Christmas spirit to the office and relaxing organizational rules and power structures (Rippin 2011).

Another powerful Christmas symbolic ritual is, of course, the giving of presents. Like paper crowns, gift giving may have its roots in the commemoration of the visit of the Magi (Rippin 2011). This ritual, which is key to family and work Christmas celebrations alike, is associated with Christmas values such as altruism and generosity, as well as showing love and appreciation for each other (Lemmergaard and Muhr 2011). Gift giving naturally results in shopping, which we explore in more depth below.

Shopping and Commercialisation

McKay (2008) argues that Christmas brings together a range of social and commercial practices that are further shaped and informed by Western capitalism and its effects on the domestic and public realm. Like many other leisure and tourism experiences, Christmas celebrations are rich in 'habitual enactments' and 'never entirely separate from the habits of everyday life' (Edensor 2001, 61). While several authors refer to the festive season as a liminal (Nash 1981) or a carnivalesque experience (Rippin 2011; Winchester and Rofe 2005), in this chapter we frame the Christmas experience as an extension, rather than simply a suspension or inversion, of the everyday. It would be almost impossible to discuss Christmas without exploring how it has become entwined with and extends consumerist lifestyles, advertising, and shopping practices. As early as 1867, Macy's first extended its Christmas shopping hours and late openings are now ubiquitous in both the United States and the United Kingdom. For Johnes (2016) the act of giving and receiving gifts was well established by the Great War but, despite now being one of the key elements of the Christmas season, it is also heavily criticised for being a consumeristic and hedonistic activity, in contrast with the 'traditional' (Christian) Christmas ethic of altruism and generosity. Some authors have argued that Christmas was reinvented in the nineteenth century specifically to support the booming mass production economies of Europe and the USA (for example, Storey 2008). Certainly, in the Victorian era the development of mass production in Europe and North America encouraged the commercialisation of gift-giving and Christmas-themed marketing after a period (1790–1836) when Christmas was hardly mentioned at all in the media (Miller 1995). The first Christmas card dates from 1843; and by the 1860s mass production of Christmas greeting cards had become common practice (Hancock and Rehn 2011). In the following half century, iconic

Christmas imagery such as Coca Cola's Santa Claus and Harrod's Christmas shop windows in London started to represent – according to Hancock and Rehn – 'a model of global economic ambition and cultural aspiration' (2011, 738).

The tension between the religious, holy meanings of Christmas and their secular, hedonistic counterparts – referred to as the paradox of Christmas (Pimlott 1962) – has characterised much academic discussion around Christmas. Many criticise the way Christmas has developed over time into a 'global festival of production and consumption' (Hancock and Rehn 2011: 737), a 'period of self-indulgence' (Rippin 2011: 830) and the 'greatest holy day of the consumer culture' (deChant 2003, 12, cited in Bartunek and Do 2011). According to Bartunek and Do (2011), commercialism has gone through a process of sacralization whereby sacred aspects of Christmas (e.g. religious hymns, charity giving) are embedded in typically commercial venues (e.g. shopping malls, advertising). This process has made commercialism 'more prominent than the religious celebration of Christmas' and 'from a commercial standpoint, Christmas is indeed the most sacred day of the year' (Bartunek and Do 2011: 803). Truzzi (1968, cited in Cluley 2011) notes the fundamental role of Santa Claus in bridging the sacred and secular realms, while Boyer (1955, 481) goes so far as claiming that Santa has replaced 'God as the figure to be worshipped at Christmas' (cited in Culey, 2011).

Before tackling these debates in relation to the specific case of London, it is worth recalling Johnes' (2016) argument that to focus only on Christmas as a time of commercialisation ignores that society in general is already commercialised. As argued earlier, the framing of Christmas as an inversion, transgression or suspension of the everyday goes some way to explaining how it has become strongly articulated with pleasure and entertainment not available at other times of the year. However, we should not ignore how Oxford, Bond and Regent Streets, across the entire year, are dedicated to shopping, entertainment, and advertising. To single out Christmas as somehow different obscures how every other time and day of the year the capital is equally embedded in capitalist relations. Similarly, Deacy argues that complaints about the commercialisation of an otherwise sacred festival imply that 'Christianity… exists in some sort of economy-free zone' (2016, 72). Equally, and while not discounting that there is a focus on commercial activity associated with Christmas, it is notable that there is an expansion of free and accessible events occurring in London at this time. Winter Wonderland (excluding the rides), the free carols in Trafalgar Square, the elaborate window displays across Bond, Regent and Oxford Streets, the New Year's Eve fireworks, as well as concerts and live entertainment are all free to the public.

As noted earlier, debates about the commercialisation of what is otherwise a religious festival have long reverberated but have perhaps become even more intense, and are now articulated with wider issues about public space and urban development. Hannigan's (1998) notion of shoppertainment becomes especially important here. In his *Fantasy City* Hannigan lays out a typology of shoppertainment districts in the United States. Some of these features are beyond the focus of this chapter, but the term does encapsulate how the Christmas period sees a synergy 'of form, content and structure as a key business strategy' (Hannigan, 1998, 63); a convergence, in other words, of shopping, entertainment, food, design and culture. In effect, Christmas shopping becomes more than just an act of buying, but also entails free entertainment in the form of shop windows, carols and lighting displays, alongside consuming roasted chestnuts from pavement sellers, as well as drinking mulled wine from nearby pubs. These otherwise distinct experiences merge into what Hannigan refers to as 'experiential retailing' (1998, 69), which finds its ultimate form in modern themed environments.

Theming in London's Public Spaces

Themed environments are spaces where there is a single narrative operating, where the dress, music, food, architectural and symbolic motifs coincide to tell a specific story or produce a singular atmosphere. Typically, we might think of such spaces as bounded, such as traditional theme or amusement parks. An early debate within the study of theme parks was precisely that they were typically walled off and distinct from their surroundings in economic, social and aesthetic terms (Moore 1980, cited in Hochbruck and Schlehe, 2010). As Hochbruck and Schlehe (2010) argue, this is not necessarily the case and themed spaces are no longer considered extraordinary or liminal; they are now integral to everyday life. In the case of London in the festive period, this is especially the case. While the period has an atmosphere of liminality and specific spaces accord with a traditional understanding of themed spaces, we might instead say that rather than being walled off, there is instead an intensification and expansion of Christmas motifs, food, dress as well as sounds, smells, and symbolic codes across different sites. Equally, this intensification typically occurs in areas already marked by a year-round commercial, touristic or entertainment purpose.

London's Christmas offer is not centralised in spatial or managerial terms, as it might be for a single event or, indeed, a themed park. There is a much more complex layering of disparate management structures, competition between different areas, and different intensities of Christmas-ness across the festive period. London is governed by a two-tier structure, the Greater London Authority (GLA) plus 32 borough councils and the City of London Corporation. The GLA is responsible for transport, strategic planning, policing and fire services amongst other tasks, while the local councils take responsibility for such everyday services as libraries, environmental health and waste disposal, as well as education, planning and social services. Both play an active role in promoting and celebrating Christmas, with local councils typically providing lighting and decorations for local high streets, while the GLA promotes and is responsible for ticketing of the New Year's Eve Fireworks Festival. Westminster Council, in which Trafalgar Square is located, manages the ceremony for the lighting of the Christmas Tree gifted annually from Norway. Local shop owners, events companies, and leisure providers also provide their own decorations and events.

Business Improvement Districts (BIDS), which were introduced to England and Wales through the Local Government Act, 2003, also play a unique role. BIDS function as representatives for local businesses within a defined geographical area and businesses within those boundaries pay a levy, which is then used to fund specific projects. The largest BID is New West End Company, which includes 600 businesses across Bond Street, Regent Street and Oxford Street in central London. Other BIDS play an equally active role in partnering with other events or charitable groups, or promoting their own Christmas related events. The BID Angel London, for example, in the city's north, hosts a market on Islington Green while also providing decorations on the main retail area, Upper Street. Baker Street Quarter Partnership, the BID north of the New West End company, also hosts their own Christmas market, capitalising on the area's food offer and partners with children's charities to collect toys for underprivileged children. Fitzrovia Partnership stage their own street lighting event, while the South Bank BID promote Christmas related performances in local theatres. What we see here is rather than a single, top-down policy for Christmas, or a single theming taking place, there is a coming together of diverse political, economic and leisure groups across multiple sites. The theming of London across the Christmas period is therefore far more dispersed and subject to multiple layers of management and indeed intentions, much more so than found in a traditional themed environment.

Other debates about theming continue to remain important, however. Much has been written from a US perspective on theming, especially in terms of how it might challenge authenticity, or condense specific histories, cultures, and spatial forms into commercialised products to be consumed (Gottddiener 1997). The commercialisation of history, and the profit motive underlying it, finds accord in Miles' argument that themed environments act as 'physical manifestations of consumer society' (2010, 142). For Bryman (2004), themed shopping malls and centres serve an important function in terms of providing an alternative to home-based internet shopping, but he remains concerned by what he refers to as 'Disneyization' creeping into wider society. As we have noted these concerns are not new, but around Christmas debates about corporatisation, commercialism, and the synthesis of specific marketing strategies become especially acute.

An example that encapsulates many of the debates raised here is Winter Wonderland. First opened in 2007 by PWR Events in conjunction with the Royal Parks, Winter Wonderland takes place at the eastern end of Hyde Park, one of London's eight Royal Parks. It is one of the largest Christmas events in the capital and features over 100 rides and attractions. While it is free to enter, the

rides on offer, which include an ice-skating rink, rollercoasters, and an observation wheel, typically cost between £3 and £8. Again, the event is not entirely
extraordinary in terms of its function or its location. It is located close to Park
Lane as well as Oxford Street, the shopping and commercial heart of London.
Hyde Park also regularly features other entertainment events, notably open-air
concerts. The event represents a winter equivalent of the summer festivals that
now regularly take over London's parks in the summer (see Chapter 10). The
emergence of other park-based events in London, including other 'park in the
dark' festive attractions such as at Kew Gardens and Chiswick House, represent
a similar trend towards capitalising on an otherwise quieter time of the tourist
calendar.

To date, Winter Wonderland has attracted over 14 million visitors and while
this would suggest it has a wide appeal, it has also attracted controversy for
the way it occupies, commercialises and is believed to 'vulgarise' a large part
of a Royal Park. As Smith (2014b) suggests, staging events in public parks has
attracted widespread controversy and raised concerns about commercialising
otherwise public land (see Chapter 10). While bearing in mind that during the
winter months the park would otherwise be closed after dusk, and thus it does
expand leisure opportunities, approximately 13 per cent of the park is given
over to this event with grassland needing to be cornered off after the event
in order for it to regrow. Winter events are particularly damaging for parks,
because the turf is less resistant to trampling and regrowth is slower, meaning
that grass has to be replanted the following spring (Smith 2016).

Winter Wonderland raises many other concerns that echo the literature on
themed environments, from its faux Bavarian market and focus on shopping to
its confusing layout. Brida et al. (2016) remind us that people do not spend as
much at Christmas markets as initially intended, suggesting either that there
is nothing available that visitors wish to buy, or that themed spaces are far less
successful in coercing people into parting with their earnings than some critics
would suggest. Nonetheless, and while it does not feature the heavy corporate
branding typically associated with themed spaces (Gottdeiner 1997), Winter
Wonderland does represent a wider process occurring across London at this
time; a knitting together of branding, marketing, and event-led strategies that
draw upon various winter and Christmas related themes and motifs. In particular, it draws heavily on the aesthetics of a German Christmas market, a trend
which has occurred across a number of markets across the capital. Christmas
Markets are not unique to London, and as Brida et al. (2016) note, in 2014 there
were over 154 Christmas markets and 2,634 smaller markets operating across
Europe. Winter Wonderland, like Winter Festival on London's South Bank,
London Bridge City Christmas Market, Christmas in Leicester Square (which
started in 2016), and Winterville on Clapham Common, all draw upon symbols of traditional Bavarian/German Christmas markets from wooden chalets
and artisan toys, to Bavarian sausages and beers. It is worth noting that these
markets do not present only a crude staging of German culture, however; vegan

pizzas and Mexican burritos are sold alongside Lebkuchen biscuits and Glüh-wein, with the music of Michael Jackson playing alongside more traditional Christmas carols.

Explaining the current fashion for Bavarian markets is complex, but Pitcher's (2014) argument in his *Consuming Race* is worth noting. He argues that the recent proliferation of 'Scandinavia cool' and all things Nordic in British culture, while not racially motivated in crude terms, represent 'ethnically appropriate' forms of consumption in a complex and global city. Spaces such as Winter Wonderland and other Christmas markets are not racially exclusive, but they are racially coded in the sense that their symbolic reference points privilege a very narrow imagining of what Christmas is and what it allegedly once was. With Christmas having become marketed through foods, music, weather patterns, and other symbols of northern Europe, the popularity of Bavarian-style markets rests on an assumed cultural and historical affinity with all things northern European. This line of argument also finds accord in Armstrong's (2008) discussion of the intricate and complex connections between German and English Christmas literature and rituals. He argues that:

> the desire for German Christmas market phenomenon indicates a desire for authenticity that reveals continuing tensions between familial or emotional expectations and consumer realities, tinged with apprehensions that something that might be described as spiritual has been lost. An investment in a 'traditional' German Christmas has the potential to allay these concerns … (Armstrong, 2008, 489).

It is not our intention here to find some original Christmas truth. Christmas is a conflation of different rituals, but the recent proliferation of festive markets with a decidedly Germanic theme does raise questions about the extent to which these events promote a less accurately cosmopolitan London, and one instead decidedly oriented towards a Christmas imagined, and appropriated, as more authentically northern European. The recent popularity of ice rinks poses similar questions, especially given the rarity of snow or freezing temperatures in the capital. While Bell (2008) frames them more in terms of entrepreneurial governance and broader cultural and leisure–led regeneration strategies, ice rinks also represent a similarly northern Europeanisation of Christmas in London.

What is perhaps most significant about the markets, fairs and ice rinks discussed here is that they clearly demonstrate the expansion of London's Christmas offer, both temporally and spatially. Parks that might otherwise be closed or areas not otherwise orientated towards tourism in the winter months become instead integral to the promotion and experience of Christmas in London. The cold and dark streets and parks of London in December are brought to life and become instead brightly lit, atmospheric and appealing. Dusk in London's winter occurs early in the afternoon, at approximately 3.30pm around the solstice, so of central importance to the allure of London at this time of year is lighting. As evidenced in the now famous Lower Morden Lane in suburban south-west London (*Time Out* 2017), elaborate lighting displays occur across the capital, including, as Maitland's chapter explores, in suburbia. As well as pointing back to the sensual, atmospheric and experiential component of Christmas celebrations, lighting and illumination also turn the mundane and relatively bleak atmosphere of London's winter streets into bright, festive spaces.

Christmas Illumination

Lights have long been a fundamental part of Christmas. The tradition of decorating Christmas trees inside the home with candles (and more recently electric illuminations) originates from Germany and is probably linked to pagan celebrations of the winter solstice, the shortest day of the year (Puiu 2016). In London, the most famous Christmas tree is the Norwegian spruce in Trafalgar Square: a twenty-metre tall tree donated by the country of Norway to London every year since 1947. In recent times, light displays on people's homes and gardens have also become popular, 'producing a particular geography of Christmas illumination' and attracting criticism from the media for being 'immodest' and 'tacky' (Edensor and Millington 2009, 104). It is public illuminations, however, that really contribute to the development of a magic Christmasscape, engaging tourists and residents alike in an aesthetic Christmas experience.

One of the first streets to ever set up a Christmas light display was Hollywood Boulevard in Los Angeles, USA in the 1920s. The street took advantage of the local cinema industry to provide an elaborate display and temporarily renamed itself 'Santa Claus Lane' (Isenstadt 2015). Visitors were astounded: 'you would blink your eyes and believe yourself transported to some other planet… in a new world, modern, splendid, gloriously illuminated with winking, colored lights' (Wilcox 1936 cited in Isenstadt 2015, 53). It took a few decades for this invented Christmas tradition to be exported to Europe. Selfridges, the iconic London department store, provided an illumination display as early as 1935 (The Guardian 2007). However, public Christmas illuminations of shopping streets did not appear in London until 1954 when the local association of retailers and businesses of Regent Street decided to provide this type of display for the first time in London (BBC News 1997; Johnes 2016). This initiative was such a success that the House of Lords tried to discipline the organisers for causing chaos and obstruction (BBC News 1997). Although this practice went through times of decline (it was, for example, interrupted in the late 1960s/70s for financial reasons), Christmas illuminations have now grown in popularity and have become a fixed feature of most British shopping streets during the festive season.

In London, the most famous and elaborate light displays are those in the West End, particularly Regent Street, Oxford Street, Covent Garden, Carnaby

Figure 9.1: Regent Street Decorated with Lanterns and Christmas Trees in 1954 (Photo: John Maltby/RIBA Collections).

Street and Bond Street. Other shopping districts provide Christmas illuminations, including Hampstead, South Kensington, South Bank, Camden, Maryle-
bone and Kingston. These are funded by local councils, as well as partnerships
between different groups, such as events companies and Business Improvement
Districts. A number of tour companies now offer guided (walking, cycling or
bus) tours of Christmas illuminations. As Linden and Linden (2016) note,
lights generate excitement and guide the (shopping) way, while a dark street
at Christmas is perceived as uninteresting and un-happening. Light displays
in non-shopping locations are less common, but in the past few years the Kew
Royal Botanic Gardens have organised a (ticketed) illumination event featuring
botanic-themed light installations, a fire show and music. This event is aimed at
drawing visitors to a prominently spring and summer attraction, thus further
extending London's offer into the colder off-season months and into the night.
Each year in November Christmas lights 'switch on' events are also staged all
over London, to signal the (early) start of the shopping season and enhance
the sense of celebration. These 'family-friendly' street festivals generally involve
music and celebrities who are invited to 'switch on' the Christmas illuminations
(Bell 2009). In 2016, Oxford Street became traffic free throughout the day for
the switch on event on 6th November. Regent Street provided a 'toy parade'
sponsored by the famous Hamleys toy shop which was allegedly attended by
750,000 people, despite the cold weather (Evening Standard 2016). Switch on
events can also be used to promote charitable causes or raise awareness of social
issues. For example, in 2016, Berwick Street in Soho used the switch on event
to raise awareness about plans to privatise the local street market (Linden and
Linden 2016).

Christmas illumination displays have been accused of being environmentally
irresponsible, certainly a reasonable argument considering, for example, that the
carbon footprint of the Oxford Street and Regent Street Christmas lights alone
would take each year over 200 trees (and 100 years) to offset (Evening Standard
2006). As a consequence of the amount of energy they consume as well as production costs, Christmas illuminations are also very expensive. Depending on
the location, costs are usually borne by local authorities, local businesses (often
through a Business Improvement District or similar partnership scheme) and
sponsors. Local authorities, however, are increasingly reducing funding due to
budget restrictions, and even local retailers have occasionally opted out due to
doubts about the actual impact on Christmas sales (Linden and Linden 2016).
Corporate sponsorships have become a popular option to help with the cost
of illuminations in shopping precincts with high footfall. Despite complaints
that sponsored lights are too commercial and vulgar, prompting attempts by
architects from the Royal Institute of British Architects (RIBA) to improve their
design (Linden and Linden 2016), corporate sponsors have now become an
accepted feature of illuminations in central London. Partnerships with charitable and cultural organisations have also been developed, including children's
charity NSPCC (Oxford Street lights in 2016), the Royal Opera House (Covent

Garden switch on event in 2016) and the Victoria and Albert Museum (Carnaby Street lights in 2016). Associations with cultural institutions and charities are perhaps more likely to be welcomed by the public, as they may be perceived as more consistent with traditional Christmas values (generosity, family, altruism) and the cultural fabric of a place.

While Christmas illuminations such as Christmas tree decorations, candles and fairy lights are a traditional feature of family Christmas rituals inside the home, what makes public street illuminations and switch on events particularly notable is their role in leveraging the Christmas spirit. They encourage a sense of celebration and create an appealing atmosphere in public spaces. Edensor's work on the role of light, darkness and light events in the creation of atmosphere (Edensor 2012; 2015) is particularly helpful here. Atmosphere is produced by a variety of tangible and intangible factors, including other people, light, sounds, architecture, sensations and representations (Edensor 2015). Research conducted in London's East End, for example, found that the presence of other people was as important as the quality of the urban environment, if not more, in the creation of an appealing atmosphere for visitors (Pappalepore et al. 2014). In this sense, atmosphere is co-created and 'prosumed' by space users. Therefore, the fact that Christmas light displays and switch on events draw visitors and residents to specific public spaces during the festive season is very important in developing that sense of celebration and conviviality sought by retailers and hospitality businesses during this crucial shopping season. But light itself is also a very powerful producer of atmosphere, as exemplified by the use of light in landscape gardens (Böhme 1993) and cathedrals of light (Edensor 2015). Illuminations enrich space with 'oneiric and phantasmagoric qualities' (Edensor 2012, 1107), thus actively contributing to the Christmas spirit atmosphere, which combines elements of dream, emotions and nostalgia (Clarke 2007).

Christmas illumination displays and Christmas light events are also important as they provide the kind of family-friendly nocturnal entertainment that can contribute to extending the urban leisure experience into the night for a wider range of audiences. This is consistent with the London Mayor's vision to enhance the city's night-time economy and improve the experience of the city at night for its users (GLA 2017). As part of this vision, a 'Night Czar' was appointed to nurture London's night-time economy and night-time transportation was improved including for the first time a 24-hour underground service. In his 2017 '24 hour London Vision', the new London Mayor highlighted the importance of decorative illuminations and the use of 'Nuit Blanches' (white nights) as tools to achieve his objectives (GLA 2017). The second edition of the Lumiere (white night) festival in London took place in January 2018, further extending the festive season to an otherwise quieter month for tourism. Light Night events – including Nuit Blanches, Lumiere festivals, late night and light art festivals – have grown in popularity around the world thanks to their potential to attract visitors, produce a festive atmosphere and change image perceptions of places in decline (Jiwa et al. 2009). In the UK, such events have also

Figure 9.2: Winter Lights – An Annual Light Festival staged at Canary Wharf every January (Photo: Andrew Smith).

been developed as part of a drive to make city centres at night more inclusive and offer a more appealing alternative to mass drinking (Evans 2012).

These night events have drawn criticism, largely for promoting a passive engagement with 'spectacle' rather than active participation in the arts (Mercer and Mayfield 2015). This view is certainly debatable in the case of light night events and Christmas events, which, on the contrary, tend to be a participative type of event often involving interactive and multi-sensory installations. If we look at illuminations as just one of the many elements contributing to the production of a Christmas atmosphere, which also include other people and conviviality, the taste of mulled wine and roasted chestnuts, and Christmas music playing in the shops, then the spectacle of lights becomes part of a multi-sensory and convivial leisure experience. The use of illumination 'has the potential to re-enchant everyday life and ordinary spaces' (Edensor, 2015: 343); spaces which, certainly in the winter months, would otherwise be less attractive for locals and tourists.

Conclusions

In this chapter we have explored the ways Christmas-themed events and spaces in London continue to expand, extend and intensify tourism and the tourist experience. The festive season, which now begins in November and extends through to January, plays a key role in drawing visitors to the capital during an otherwise slow season. Thanks to the proliferation of newly created Christmas rituals such as Black Friday, the lighting 'switch on events', Diwali and Lumiere markets and light festivals, the duration of the Christmas season has now been extended to last well beyond December. Alongside the usual Dickens stories, modern romantic Christmas films such as *Love Actually* (2003) and *The Holiday* (2006) have also played a role in attracting international visitors keen to experience London's Christmas atmosphere. In addition to drawing more visitors to London, Christmas contributes to extending the times when residents and tourists experience the city. Temporary themed spaces such as Christmas markets, ice rinks and night illuminations all provide entertainment into the evening: a time, particularly during the winter season, when London would be typically dominated by less family friendly leisure activities such as dining out, drinking and clubbing. Similarly, special events such as Christmas at Kew in the Royal Kew Gardens and Winter Wonderland in Hyde Park provide themed tourist attractions in green spaces that would otherwise be closed to the public at night.

Christmas in London offers expanding opportunities for leisure and sociability in what otherwise would be the quieter months but 'how' the festive period is celebrated has attracted controversy. The way in which Christmas values such as generosity, gift giving and conviviality are leveraged by commercial businesses to increase sales is widely criticised for turning a holy event into a 'global festival of production and consumption' (Hancock and Rehn 2011, 737). Similar to other large events (see Chalip 2006; Smith 2014a), leveraging tactics such as facilitating sociability, producing related ancillary events, and theming are used to leverage the emotional and symbolic power of Christmas. Christmas illuminations for instance, which were created with the intent of encouraging spending in the first place, are now being designed to meet the specifications of sponsors. Another consequence of the commercialisation of Christmas is the proliferation of Christmas-themed markets in public spaces, which is seen by some as an unwelcome occupation of public space (thus reducing rather than extending visitors' and residents' public space experiences in the city). However, contemporary society is already commercialised, and although criticism is important, a more comprehensive analysis of this complex global phenomenon is needed. As alluded to in our introduction, London's great strength, its diversity and multiculturalism, has shaped and informed the ways the festive months are now celebrated and marked. The Diwali festival to the Chinese New Year, and all that occurs in between, are now a fundamental part of the city's offer. And though, as noted above, the coding of Christmas via uniquely northern European symbols and codes at Winter Wonderland and other Christmas Markets perhaps does not do justice to London's diversity, they are components of a greater whole where the secular, sacred, commercial and free events come together. Christmas traditions, far from having solely religious origins, have always comprised pagan and Christian rituals, symbols and behaviours. And while commercial aspects have certainly intensified in the last century, spiritual aspects such as giving, spending time with family, self-reflection and religious rituals are still important. Meanwhile, creative activities such as the production of Christmas plays and costumes, cooking and crafting of decorations, may contribute to enhancing a form of existential authenticity (Wang 1999), thus further contributing to enhancing the Christmas experience. Themed spaces and events, rather than providing a suspension of the everyday, facilitate the enhancement of everyday rituals such as conviviality, shopping and a multi-sensory, multi-sensual experience of the city. Many leveraging tactics seen as merely commercial and exploitative – such as Christmas markets, illuminations and ancillary events – also provide free entertainment and contribute to the development of a Christmas atmosphere. The expansion and intensification of Christmas in London has therefore led to some new, and some rather more dated, concerns, but it has also enabled the expansion of the tourist season well into otherwise quieter months and added to the capital's tourism offer.

Notes

[1] Non-routine leisure day visits of at least 3 hours. Data available from https://gbdayvisitslightengland.kantar.com/ViewTable.aspx. Data for 2016 was not available at the time of writing.

[2] Data from the Great Britain Visits survey 2015, available from https://www.visitbritain.org/online-data-browser. Data for overseas visitors is only available quarterly for London. 2016 data was not yet available at the time of writing.

References

Armstrong, Neil. 2008. 'England and German Christmas Festlichkeit, c. 1800–1914'. *German History*, 26(4), 486–503.

Bartunek, Jean and Do Boram Do. 2011. 'The Sacralization of Christmas Commerce'. *Organization*, 18(6), 795–806.

Belk, Russell. 2001. 'Materialism and the Making of the Modern American Christmas'. In Daniel Miller, ed, *Consumption: Critical Concepts in the Social Sciences*. London: Routledge.

Bell, David. 2009. 'Winter Wonderlands: Public Outdoor Ice Rinks, Entrepreneurial Display and Festive Socialities in UK Cities'. *Leisure Studies*, 28(1), 3–18.

Böhme, Gernot. 1993. 'Atmosphere as the Fundamental Concept of a New Aesthetics'. *Thesis Eleven* 36(1), 113–126.

Boyer, Bryce. 1955. 'Christmas "Neurosis"'. *Journal of the American Psychoanalytic Association*, 3(3), 467–488.

Brida, Juan Gabriel, Meleddu, Marta, and Oksana Tokarchuk. 2017. 'Use Value of Cultural Events: The Case of the Christmas Markets'. *Tourism Management*, 59, 67–75.

BBC News. 1997. 'Special Report: Shedding Light on Christmas'. *BBC News*. Available at http://news.bbc.co.uk/1/hi/special_report/for_christmas/_new_year/christmas_decorations/39383.stm (accessed 6 September 2017).

BBC News. 2015. 'Christmas Conceptions lead to Autumn Baby Boom'. *BBC News*. Available at http://www.bbc.co.uk/news/health-35131097 (accessed 1 September 2017).

Bryman, Alan. 2004. *The Disneyization of Society*. London: Sage.

Chalip, Laurence. 2004. 'Beyond Impact: A General Model for Sport Event Leverage'. In Brent Ritchie and Daryl Adair, eds, *Sport Tourism: Interrelationships, Impacts and Issues*. Clevedon: Channel View Publications.

Chalip, Laurence. 2006. 'Towards Social Leverage of Sport Events'. *Journal of Sport and Tourism*, 11(2), 109–127.

Clarke, Peter. 2007. 'A Measure for Christmas Spirit'. *Journal of Consumer Marketing*, 24(1), 8–17.

Cluley, Robert. 2011. 'The Organization of Santa: Fetishism, Ambivalence and Narcissism'. *Organization*, 18(6), 779–794.

Cohen, Erik. 1988. 'Authenticity and Commoditization in Tourism'. *Annals of Tourism Research*, 15(3), 371–386.

Connelly, Mark. 2012. *Christmas: A History*. London: I. B. Tauris.

Crouch, David and Luke Desforges. 2003. 'The Sensuous in the Tourist Encounter: Introduction: The Power of the Body in Tourist Studies'. *Tourist Studies* 3(1), 5–22.

Davis, Derek, ed. 2010. *The Oxford Handbook of Church and State in the United States*. Oxford: Oxford University Press.

Deacy, Christopher. 2016. *Christmas as Religion: Rethinking Santa, the Sacred and the Secular*. Oxford: Oxford University Press.

DeChant, Dell. 2003. 'The Economy as Religion: The Dynamics of Consumer Culture'. *The Civic Arts Review*, 16(2), 4–12.

Ebbensgaard, Casper Laing. 2015. 'Illuminights: A Sensory Study of Illuminated Urban Environments in Copenhagen'. *Space and Culture*, 18(2), 112–131.

Edensor, Tim. 2001. 'Performing Tourism, Staging Tourism: (Re) Producing Tourist Space and Practice'. *Tourist Studies*, 1(1), 2001, 59–81.

Edensor, Tim. 2012. 'Illuminated Atmospheres: Anticipating and Reproducing the Flow of Affective Experience in Blackpool'. *Environment and Planning D: Society and Space*, 30(6), 1103–1122.

Edensor, Tim. 2015. 'Light Design and Atmosphere'. *Visual Communication* 14(3), 331–350.

Edensor, Tim and Steve Millington. 2009. 'Illuminations, Class Identities and the Contested Landscapes of Christmas'. *Sociology* , 43(1), 103–121.

Evans, Graeme. 2012. 'Hold Back the Night: Nuit Blanche and All-Night Events in Capital Cities'. *Current Issues in Tourism,* 15(1–2), 35–49.

Evening Standard. 2006. 'Put that Christmas Light Out'. *Evening Standard.* Available at https://www.standard.co.uk/news/put-that-christmas-light-out-7083119.html (accessed 9 September 2017).

Evening Standard. 2016. 'Thousands of Excited Children Lined Regent Street to see their Favourite Characters Come to Life at the Hamleys Christmas Toy Parade'. Available at https://www.standard.co.uk/news/london/hamleys-christmas-toy-parade-thousands-gather-to-watch-largerthanlife-characters-ride-through-regent-a3400206.html (accessed 7 September 2017).

Feldman, Stephen. 1998. *Please Don't Wish me a Merry Christmas: A Critical History of the Separation of Church and State.* New York: NYU Press.

Gottdiener, Mark. 1997. *The Theming of America: Dreams, Visions, and Commercial Spaces.* Boulder, CO: Westview Press.

Guardian, The. 2007. 'Seventy Years of West End Christmas Lights'. *The Guardian,* 14 December. Available at https://www.theguardian.com/uk/gallery/2007/dec/14/christmas (accessed on 7 September 2017).

Hancock, Philip and Alf Rehn. 2011. 'Organizing Christmas'. *Organization* 18(6), 737–745.

Hannigan, John. 1998. *Fantasy City: Pleasure and Profit in the Postmodern Metropolis.* London: Routledge.

Hochbruck, Wolfgang and Judith Schlehe. 2010. 'Introduction: Staging the Past'. In Judith Schlehe, Michiko Uile-Bormann, Carolyn Oesterle and Wolfgang Hochbruck, eds, *Staging the Past: Themed Environments in Transcultural Perspectives* New Brunswick, NJ and London: Transaction Publishers.

Isenstadt, Sandy. 2015. 'Los Angeles'. In Sandy Isenstadt, Margaret Maile Petty and Dietrich Neumann, eds, *Cities of Light: Two Centuries of Urban Illumination.* London: Routledge.

Johnes, Martin. 2016. *Christmas and the British: A Modern History.* London: Bloomsbury.

Jokinen, Eeva, and Soile Veijola. 1994. 'The Body in Tourism'. Theory, *Culture and Society* 11 (3): 125–151.

Lemmergaard, Jeanette and Sara Louise Muhr. 2011. 'Regarding Gifts—on Christmas Gift Exchange and Asymmetrical Business Relations'. *Organization,* 18(6), 763–777.

Lindem Henrik and Sara Linden. 2016. 'Switched-on City: How London Learned to Love Christmas Lights'. *The Conversation,* 22 December. Available at https://theconversation.com/switched-on-city-how-london-learned-to-love-christmas-lights-70609 (accessed 26 January 2018).

Makulski, Beatrice. 2015. *The Impact of Population Mobility and Multiculturalism on Celebratory Events, Case Study: Christmas in the UK*. Unpublished MA Dissertation. London: University of Westminster.

McKay, George. 2008. 'Consumption, Coca-colonisation, Cultural Resistance – and Santa Claus'. In Sheila Whiteley, ed *Christmas, Ideology and Popular Culture*. Edinburgh: Edinburgh University Press.

Mercer, David and Prashanti Mayfield. 2015. 'City of the Spectacle: White Night Melbourne and the Politics of Public Space'. *Australian Geographer*, 46(4), 507–534.

Miles, Steven. 2010. *Spaces for Consumption*. London: Sage.

Miller, Daniel. 1995. 'A Theory of Christmas'. In Daniel Miller, ed *Unwrapping Christmas*. Oxford, Oxford University Press.

Nash, Jeffrey. 1981. 'Relations in Frozen Places: Observations on Winter Public Order'. *Qualitative Sociology*, 4(3), 229–243.

NWEC. 2016. *West End: Christmas Tracker Report 2016*. London: New West End Company.

Nissenbaum, Stephen. 1996. *The Battle for Christmas*. New York, NY: Knopf.

Pappalepore, Ilaria, Maitland, Robert, and Andrew Smith. 2014. 'Prosuming Creative Urban Areas. Evidence from East London'. *Annals of Tourism Research*, 44, 227–240.

Pimlott, J. A. R. 1962. 'But Once in a Year'. *New Society*, 12(19), 9–12.

Pitcher, Ben. 2014. *Consuming Race*. London: Routledge.

Pretes, Michael. 1995. 'Postmodern Tourism: The Santa Claus Industry'. *Annals of Tourism Research*, 22(1), 1–15.

Puiu, Tibi. 2016. '*The Origin of the Christmas Tree: From Paganism to Modern Ubiquity*'. Available at http://www.zmescience.com/science/history-science/origin-christmas-tree-pagan/(accessed on 6 September 2017).

Rippin, Ann. 2011. 'Ritualized Christmas Headgear or "Pass me the Tinsel, Mother: It's the Office Party Tonight"'. *Organization*, 18(6), 823–832.

Rook, Dennis. 1985. 'The Ritual Dimension of Consumer Behavior'. *Journal of Consumer Research*, 12(3), 251–264.

Smith, Andrew. 2014a. 'Leveraging Sport Mega-Events: New Model or Convenient Justification?'. *Journal of Policy Research in Tourism, Leisure and Events*, 6(1), 15–30.

Smith, Andrew. 2014b. 'Borrowing Public Space to Stage Major Events: The Greenwich Park Controversy'. *Urban Studies*, 51(2), 247–263.

Smith, Andrew. 2016. *Events in the City: Using Public Spaces as Event Venues*. London: Routledge.

Storey, John. 2008. 'The Invention of the English Christmas'. In Sheila Whitely, ed, *Christmas, Ideology and Popular Culture*. Edinburgh: Edinburgh University Press.

Time Out. 2017. 'Meet the Residents who Live on London's Most Christmassy Street'. *Time Out*. Available at hhtp://www.timeout.com/london/news/

meet-the-residents-who-live-on-londons-most-christmassy-street-121917 (accessed 18 January 2018).

Truzzi, Marcello. 1968. *Sociology and Everyday Life*. Englewood Cliffs, NJ: Prentice-Hall, Inc.

Veijola, Soile and Eeva Jokinen. 1994. 'The Body in Tourism'. *Theory, Culture and Society,* 11(3), 125–151.

Wang, Ning. 1999. 'Rethinking Authenticity in Tourism Experience'. *Annals of Tourism Research,* 26(2), 349–370.

Wilcox, G. 1936. 'Christmas Comes to Hollywood'. *The Atlanta Constitution,* 20 December, 28.

Winchester, Hilary and Matthew Rofe. 2005. 'Christmas in the "Valley of Praise": Intersections of the Rural Idyll, Heritage and Community in Lobethal, South Australia'. *Journal of Rural Studies,* 21(3), 265–279.

CHAPTER 10

Event Takeover? The Commercialisation of London's Parks

Andrew Smith

Introduction

Over the past forty years, public parks have had to endure periods of underfunding which have instigated inefficient cycles of decline and regeneration (Smith et al. 2014). These 'crises' are significant for various reasons, not least because they are usually accompanied by changes in the way parks are conceived, managed and governed. As Krisinsky and Simonet (2012) argue, park crises such as the one experienced in the US in the 1970s tend to be used as an excuse to justify further private sector involvement. They are an intrinsic part of the neoliberalisation project which is known to function through processes of creative destruction (Brenner and Theodore 2002). The austerity policies pursued by the UK government since 2010 mean many UK parks are currently experiencing the latest funding crisis. Local authorities have had to cope with significant cuts, with park budgets hit particularly hard: 92 per cent of UK park managers reported reductions in their budgets in 2013–2015 with 33 per cent experiencing cuts of over 20 per cent in the same period (Heritage Lottery Fund 2016).

How to cite this book chapter:
Smith, A. 2019. Event Takeover? The Commercialisation of London's Parks. In: Smith, A. and Graham, A. (eds.) *Destination London: The Expansion of the Visitor Economy.* Pp. 205–223. London: University of Westminster Press. DOI: https://doi. org/10.16997/book35.j. License: CC-BY-NC-ND

The financial crisis currently facing parks provides the context for this chapter which explores the implications of the push for new sources of revenue. Parks have long earned revenue from concessions and charges, but many are now required to generate a substantial proportion of their funding via commercial income streams. This is changing the way parks are funded, but it is also transforming the ways they are governed and managed. Government austerity combined with the ongoing project to neoliberalise public services is creating a new breed of 'entrepreneurial parks' (Davidson 2013) which are 'financially self-sustaining' (Loughran 2014), an ambition which is assisted 'by having parks managed and maintained by private companies' (Davidson 2013: 657). In London, this trend is one that is strongly resisted both by local users (e.g. Park Friends groups) and by campaign groups (e.g. The Open Spaces Society) seeking to protect public access to open spaces.

The crisis facing UK parks, and the commercialisation likely to result from it, are national challenges rather than ones specific to London. But these issues are particularly relevant to the UK capital. London is a city famed for its parks, and as the permanent and temporary populations of the city have grown, these spaces are heavily used as sites for everyday recreation and tourist visitation. However, London's parks are not merely places for people to use and visit, they are highly symbolic sites that are coveted by various political and commercial interests. These multiple roles are often incompatible. The commercial exploitation of London's parks interferes with everyday use by restricting access to space and by encouraging certain types of user. Commercialisation also affects the established role of prominent green spaces like Hyde Park, Jubilee Gardens and Blackheath as sites of political gatherings and resistance. The analysis here highlights that London's parks are inherently contested spaces because the territorial demands of citizenry, capital and state collide.

This chapter examines the ways that London's parks are increasingly commercialised and it explores the various issues associated with this trend. These include a number of ideological concerns alongside problems with the efficacy of commercial funding. The paper then focuses on one of the most obvious and prevalent ways that London's park authorities are generating income – by staging commercial events. Events provide insightful examples of the way that public parks are being privatised in subtle and incremental ways. Hiring out parks to organisers of festivals, exhibitions and sports events provides a way of generating income from park space without having to sell off public assets. Indeed, the rise of park events illustrates how London's public spaces are increasingly offered for hire to private companies, something that erodes their public status. Commercial events involve temporary incursions, but their temporal footprint extends well beyond the duration of events and, combined with the potential for events to act as precedents for other further commercialisation, this means the rise of event funding has significant implications for the accessibility of parks.

The chapter begins with an overview of park commercialisation which is used to contextualise the subsequent discussion about London's parks' exploitation as event venues. In several instances, the introduction of commercial events into London's parks has been strongly resisted by local users. These disputes came to a head in 2016 when campaign groups launched legal actions against large-scale events in Battersea Park and Finsbury Park. These cases are discussed here alongside other insightful examples where local authorities have prioritised events as a way to help pay for parks. The paper is based on a series of research exercises undertaken in the summer of 2016 including interviews with park stakeholders, extensive observation exercises before, during and after events, and online communication with park users.

The Neoliberalisation and Commercialisation of Parks

The commercialisation of parks can only be understood in the wider context of neoliberalisation of urban space. Within the enormous amount of published work on this theme, there are several texts that specifically examine neoliberal transformations of parks. For example, Krinsky and Simonet (2011) discuss the changes to staffing arrangements in New York City's parks, highlighting the increased use of non-unionised staff working for private contractors. In the contemporary era, staffing costs are also reduced by using volunteer labour – a noted characteristic of parks governed by neoliberal regimes (Rosol, 2010). Other texts focus on the transfer of responsibility from public to private organisations. For example, using Harvey's (1989) conceptualisation, Perkins (2009; 2616) examines how 'cash strapped and/or fiscally conservative local governments unload them in what amounts to a shift from state managerialism to entrepreneurial regimes of governance'. New management arrangements take various forms, including not for profits, social enterprises, private trusts, private companies, and various versions of Business Improvement Districts. Parks have always generated money from concessions, but in some neoliberal regimes, companies are not only invited to operate in parks, but to manage them as well. Perkins (2009) examines the ways some US parks have been leased to coffee chains who assume responsibility for maintenance.

Whilst many of the initiatives discussed above aim to address funding shortfalls by reducing maintenance costs, neoliberalism is also associated with various efforts to generate more income from parks. Work by Zukin (1995) and Madden (2010) in New York City shows how Bryant Park pioneered the introduction of commerce 'into what was previously the non-commercial domain of the municipal parks' (Madden 2010, 188). Loughran (2014) analyses how this has been accomplished in New York's newest park – the High Line – through the way that the space is structured and controlled – with commerce and consumption prioritised in the design and regulation of the park. This case – regarded by Loughran (2014, 50) as 'an archetypal urban park of the neoliberal era' – also provides an illustrative example of the way parks are increasingly justified, created and maintained via the value they add to surrounding real estate (Millington 2015).

Attempts to generate more commercial revenue from parks take various forms. One option is to attract sponsorship/advertising, with companies purchasing the right to be associated with whole parks or specific features within them. In extreme cases this has involved selling the naming rights to parks. A second option is to lease space to commercial enterprises, with park authorities earning income through ground rent and/or a levy on ticket sales. This model encompasses both semi-permanent installations (e.g. visitor attractions) and temporary ones (e.g. events). For example, several London parks (Alexandra Palace Park, Trent Park and Battersea Park) now feature Go Ape attractions which require entry fees to access installations installed above ground in the trees (see Chapter 6). Park authorities can also generate income directly via introducing/raising charges for certain services (e.g. educational courses), permits (e.g. for commercial photography) and licenses (e.g. for fitness training). Parks have also attempted to increase the scope of charges levied for sport facilities by introducing fees for those wishing to play organised sport. For example, Regents Park in London now charges people who want to play football or cricket in areas designated for organised sport.

The sorts of commercial activities outlined above can generate a significant amount of revenue, particularly for iconic parks located in large cities. The Royal Parks – the agency responsible for eight historic parks in London – have pioneered the shift towards commercial funding. This was a transition they were required to make to offset reductions in their annual government grants. In 2015/6, the cost of managing the Royal Parks was £34.9 million and 64 per cent of these funds were raised via commercial income – through a mixture of events, sponsorship, donations, catering, grants, lottery funding, licences, and rental income from lodges, filming and photography (The Royal Parks 2016). A shift towards more commercial funding has been controversial and some of the new initiatives have been vehemently opposed. For example, in 2014 campaigners forced The Royal Parks to drop their plans to charge people who regularly played softball in Hyde Park.

Park commercialisation in the UK has been pushed by several quasi non-governmental organisations (QUANGOs) that, in the absence of national government involvement in parks, have assumed key leadership roles. The Heritage Lottery Fund (HLF) – established in 1996 to distribute money generated through the UK National Lottery – has become particularly influential given they are one of the few organisations with the resources to fund park projects. The HLF have co-produced reports and toolkits that encourage parks to develop their commercial potential. For example, in 2011 the HLF, the Big Lottery Fund and The Land Trust launched 'Prosperous Parks', an income generation toolkit which helps parks to generate ideas for expanding their commercial activity. Several London parks – including Battersea Park and The Royal Parks – are cited in Prosperous Parks as exemplars of commercialisation which other park authorities are encouraged to emulate. The HLF supported by Nesta (the UK's innovation foundation) also commissioned the Rethinking Parks project which tested 'new business models for parks in the twenty-first century' (Nesta 2013; 2016). The neoliberal/commercial emphasis of these initiatives was justified through the notion that new approaches were needed to ensure that public parks would 'remain free, open and valued community assets' (Nesta, 2016, 4). Citing the threat that parks will be closed or sold off unless new financing models are adopted is a common way of legitimising the commercialisation of parks.

There are understandable concerns about the explicit commercialisation of public parks being pursued in the UK and in other neoliberal regimes. These reflect wider anxieties about the commercialisation of urban public spaces (Kohn 2004). Introducing commercial activity means that parks are increasingly oriented towards consumers, with those unwilling or unable to pay unfairly excluded (Madden 2010). Commercialisation sits awkwardly with the history and ethos of public parks as open, accessible spaces that are free to use by anyone. Even small increases in the amount of commercial activity provide precedents for more extensive commercialisation, laying the foundation for more controversial changes – including charging for entry. The greater involvement of private companies in public parks – through sponsorship, product launches and entertainment facilities – erodes their historic function as places that are ostensibly different from the rest of the built environment that surrounds them. And whilst it may be unrealistic to keep the commercialisation of the contemporary city out of public parks, that does not mean we should not try to.

Concerns about commercialisation extend beyond the predictable critique of the denigration of public space: there are also significant issues with the efficacy of commercial funding. For example, whilst commercialisation is justified on the basis that funds are needed to pay for parks, in many instances the revenue earned is not hypothecated and spent on specific spaces, or even parks in general. Instead it goes into general budgets, leading to the accusation that parks are becoming lucrative cash cows for local authorities desperately seeking funds. The lack of transparency about where park revenue goes undermines the rationale for commercial funding. Even when cash is ring-fenced to be spent on the park in which it is earned this creates potential problems. Some parks are better positioned to capitalise on opportunities to generate commercial revenue and these tend to be those surrounded by affluent communities. Therefore, the rise of commercial funding exacerbates existing inequalities in the provision of urban green space (Millington 2015).

The drive for commercialisation is one of the key reasons why responsibility for parks is increasingly being detached from local authorities. To provide incentives to generate more revenue, and to allow that revenue to be spent on maintaining specific parks, new governance arrangements have been conceived. In some cases, these involve new autonomous or semi-autonomous agencies being established to run individual parks (or a set of parks) on behalf of local authorities. In other instances, partnerships have been established with private companies, with specialised expertise used to increase commercial revenue. For example, the London Borough of Bromley recently appointed a commercial manager for Crystal Palace Park with profits generated by new projects shared 50:50 with the Council. Commercial expertise is something many park authorities do not have, and even when parks are able to employ commercially savvy staff, there are fears that these roles are now being prioritised at the expense of other skills (e.g. horticulture).

Whilst academic texts tend to be highly critical of park commercialisation, this view is not necessarily shared among the general public. A recent survey of park users in the UK (n=2,130) revealed that 75 per cent supported more sponsorship by businesses and 59 per cent supported more commercial use (HLF 2016). Commercialisation can achieve more than just financial returns with some anecdotal evidence suggesting the presence of commercial services can make parks feel safer, add to the range of facilities on offer and diversify the profiles of users (Zukin 1995). This suggests that there may be potential for a more progressive commercialisation agenda that aims to achieve more than just revenue.

Commercial Events

The open space provided by urban parks mean they have always been earmarked as places to stage events. Hyde Park famously hosted the Great Exhibition in 1851 and ever since London's parks have been used to stage a range of events: civic occasions, political rallies, sport events, concerts and exhibitions. However, in recent years, London's parks have been more intensively used for larger, more commercially oriented events. Evidence provided by London and Partners to the London Assembly's Environment Committee suggests that commercial events have increased by 20 per cent in the last two years, with the fastest growth being in major events attended by 5,000–50,000 people (London Assembly 2017). Analysing specific examples of parks reveals even more dramatic growth. For example, in 1991 Battersea Park staged approximately 100 events (Wandsworth Borough News 1991), but twenty-five years later there were estimated to be over 600 events staged there every year (Interview with Wandsworth Borough Council [in 2016]). Many of these are small-scale community or charity events, but the extent of growth illustrates that London's parks are now perceived, used and licensed as event venues.

The growth in park events reflects growth in the events sector more generally, with experiences growing in popularity at the expense of the consumption of material goods (Pine and Gilmore 1999). The increased popularity of music festivals – and the introduction of new festivals in urban locations – has also contributed to the growth of large-scale commercial events in London's parks. However, there are other factors driving this trend, most notably the need for local authorities to generate income to pay for their parks and other services. Park authorities generate income from events in various ways: by organising their own events and charging for tickets and trading licenses or, as is increasingly the case, by hiring their spaces out to events companies and taking a proportion of the ticket sales. There is even potential to generate income by offering 'pre-event advice services for private events on council land' (Lewisham Council 2016).

Many local authorities in London have set ambitious targets to grow the amount of income they earn from commercial events. Brent Council are 'exploring the potential to hold large scale events in parks aiming for audiences at a minimum level of 2000' to fulfil their target of generating £650,000 from festivals and events in 2017/8 (Brent Council 2017). Lambeth Council has even more ambitious targets – aiming to generate £1.5 million per annum from events staged in five key locations – four of which are parks or green spaces (Event Lambeth 2015). To ensure host parks benefit the Council plans to introduce a Parks Investment Levy which will be charged to each event staged – with commercial events charged 50p per person per day. This replaces the system used in many parks which requires event organisers to pay an environmental impact fee – with these funds directed to park budgets.

London boroughs can earn significant sums from staging single events in their parks. Formula E paid Wandsworth Council £1 million for each weekend of motor racing they staged in Battersea Park in 2015 and 2016 – perhaps the most lucrative events ever staged in a London park. It costs approximately £3.25 million a year to run Battersea Park, so staging the events covered almost a third of the annual budget (Interview with Wandsworth Council [in 2016]). Wandsworth Council do not hypothecate revenue, so the money was not directly allocated to Battersea Park. However, to placate opposition, the Council promised that 20 per cent of the revenue earned would be spent on making specific improvements to the host Park. Because they were worried about being undercut by other London boroughs, and because Formula E were worried about offending other hosts who were not getting such a generous deal, the Council did not initially reveal how much the contract with Formula E was worth (Interview with Wandsworth Council [in 2016]). Councils are often reluctant to disclose how much they are being paid by event companies but this lack of transparency often breeds suspicion amongst local residents.

The fees each year paid by Festival Republic – organisers of the Wireless Festival – to Haringey Council also represent a significant proportion of the local authority's parks budget. For the 2016 edition, £446,264 in fees was generated by this three-day event (Haringey Borough Council, 2016) and, according to Haringey Council, the income from Wireless is spent on maintaining and improving the park. However, critics suggest that rather than supplementing

Figure 10.1: Battersea Park hosted controversial Formula E Motor Races in 2015 and 2016 (Photo: Andrew Smith).

the parks budget, this income is merely offsetting ongoing cuts. Staging these events provides a justification to reduce funding for parks, leading to a situation where parks have become reliant on precarious commercial income, rather than public funds. The Chair of the London Friends of Green Spaces Network suggests, that the attitude of Haringey Council is:

> if Finsbury Park is generating £700,000 or £800,000 a year, we'll take that money off the core budget for the entire park service. And to be honest, that is more than 50 per cent of the budget. So what they've done is effectively mortgaged the entire park service to be totally dependent on the commercial concerts in Finsbury Park. (Interview with Chair of London Friends of GPN [in 2016])

Events are seen as particularly attractive ways of generating revenue for parks because they can deliver wider benefits too. They add to the range of attractions that parks offer, bringing in new uses and diversifying the profile of people who visit. However, there is little hard evidence that events do diversify park users, and when new users are attracted, they tend to be those willing and able to spend (Interview with Parks Alliance [in 2016]). There are also longer-term benefits via the promotional effects of events that are represented in a range of traditional and social media. In an era of place marketing, events are seen as useful ways of promoting parks and enhancing the ways that they are perceived by target audiences. A recent report published by the London Assembly (2017) suggested that one of the key challenges facing some of London's parks is their lack of visibility. Events provide an obvious way of addressing this challenge. The need to be more visible is related to wider commercialisation as parks now need to drive demand for commercial services and attractions. There is also a danger of aestheticisation, the 'superficial embellishment of public space into visually appealing lifestyle amenities and domains of experience' – something that breeds exclusion (Glover 2015, 104).

As the recent Parliamentary Inquiry into Public Parks revealed, the increased use of parks as venues for commercial events has been met with a great deal of resistance. The amount of time and space occupied by these events is deemed to be inappropriate by other users. Events interrupt the everyday use of parks and installations often take a long time to set up and take down, causing significant disruption. The presence of large crowds and heavy vehicle movements cause damage to turf (particularly after wet weather) and this can mean park environments are inaccessible for long periods of time while they are regenerated. The weekend long SW4 music festival staged on London's Clapham Common in 2014 provides an illustrative example. The set-up of this festival began on 18 August, but due to the extended time it took to repair the damaged site, fences were not removed until 23 October. This type of disruption often takes place in the summer months when parks are most heavily used – maximising the displacement of everyday users.

Alongside concerns about the environmental damage and disruption to access, there are also significant ideological issues associated with events. Staging ticketed events commercialises park space in several interrelated ways: by turning parks into commodities that are offered for hire; by introducing charges to access parks; by normalising the presence of commercial vendors; and by providing platforms for sponsorship that wouldn't otherwise be permissible. Therefore, whilst events involve temporary installations, they have enduring effects on the ways our public parks are conceived and experienced, including material legacies. Opponents have suggested that some of the physical changes made to parks to allow them to stage events have been deliberately retained to facilitate further commercialisation. For example, the Battersea Park Action Group suggested that the extra tarmac laid down to stage the Formula E Grand Prix in 2015 was retained to allow the park to accommodate film trailers. The failure to restore the original gates to Greenwich Park after it staged the Olympic equestrian events was also seen by local campaigners as a change designed to facilitate the lorry movements needed to stage future events (Smith 2014).

Whilst complaints about events staged in London parks are nothing new, as the number and size of events has grown, opposition has intensified. In some instances, for example in Battersea Park (in Wandsworth) and Finsbury Park

Figure 10.2: The London Parks that host Major Events (mainly Music Festivals) every Summer (© Mason Edwards).

(Haringey), resistance has focused on specific events deemed to be inappropriate. In 2015 and 2016, Battersea Park hosted Formula E motor races which were vehemently resisted by the Battersea Park Action Group, whose campaigning eventually resulted in the races being discontinued. In Finsbury Park the Friends group have campaigned for several years against Wireless – a three-day music festival that is staged every July. In other instances, for example in Gunnersbury Park (Hounslow/Ealing) and Victoria Park (Tower Hamlets/Hackney), ongoing complaints about events are based more on the regularity with which they are staged. Opposition to events tends to be dominated by local concerns over access, disruption, noise and damage, but more ideologically driven resistance to the event driven commercialisation of urban parks is emerging. For example, in December 2016 the Open Spaces Society – the UK's oldest conservation body – launched its 'Save our Spaces' campaign to tackle the 'abuse' of parks in England and Wales. They cited events in various London boroughs as egregious examples of inappropriate commercial exploitation.

Victoria Park

Staging commercial events in London's public parks divides opinion. Whilst this trend has generated a lot of opposition, there are also many people who like the opportunity to attend festivals, events and exhibitions in their local park. The way users feel about park events is explored in more detail here through the case of Lovebox – a music festival that was staged in Victoria Park in East London every summer until 2017. Just before the 2016 occurrence of this event an article written by the author about park events (entitled: 'Is it right to use public parks for commercial events?') was posted on The Friends of Victoria Park Facebook site. This provoked a large number of interactions and these are analysed below to help understand the different views about this event.

Lovebox is a two-day music festival staged every July which has become a favourite haunt for London's cool hipster crowd. The event was held in Victoria Park from 2005–2017, regularly attracting crowds of 40,000 people per day – making it one of London's largest music festivals. This is an expensive event – tickets for the weekend cost over £100. The organisers, Mama Festivals – part of the Live Nation Entertainment company which dominates the music festival market – paid Tower Hamlets Council around £300,000 every year for the rights to use Victoria Park and some of this money was used to fund park improvements. Posts to the Friends of Victoria Park Facebook group suggested that opinion about Lovebox amongst local residents was divided 50/50.

Most opponents of Lovebox were not against staging events in general; they just felt this particular event was too big for Victoria Park. They also felt the event occupied too much time: 'a two day festival is actually more like a month's festival as there's the build-up, break down and recovery of [the] damaged area'. Unsurprisingly, the noise of the event was the cause of several

Figure 10.3: Lovebox – One of London's biggest Music Festivals now staged in Gunnersbury Park (Photo: Andrew Smith).

complaints, with one resident moaning that 'the music vibrates my windows and walls'. Others were concerned about disruptive effects on 'the peace of the park' and worries were expressed about the likely impact on wildlife and 'the natural rhythms of nature'. Some people felt that too much of the money earned went to the festival organisers not the Council, which led one respondent to recoil at the 'profiteering' involved. A recurring theme was the strong dislike of the oppressive structures used to fence off the festival site which were described as 'hostile and aggressive' by one contributor and as 'prison walls' by another. The sentiments of those who opposed staging Lovebox in Victoria Park were summed up neatly by one contributor who simply stated: 'I'm for letting parks be parks'.

Despite the resistance to Lovebox expressed by many contributors to the Friends of Victoria Park Facebook group, roughly an equal amount of people who responded to the post supported the event, mainly because of the money it generated for the Council and the improvements to the Park that had been made as a result. Negative impacts were acknowledged, but many felt the inconvenience and disruption were limited to a few weekends of the year – so were justified. The wider economic impact was also cited when people justified their support: 'It raises the profile of our area and provides a boost to local businesses'. Advocates felt that the events provided local residents with convenient access to great music and generated valuable experiences for the young people who attended. One contributor even felt the event contributed a (rare) feeling of togetherness amongst park users: 'even in the park so many of our experiences seem to be fairly solitary and almost in spite of one another (cyclists vs dog walkers etc). Festivals, free and paid for, give us the chance to be a bit collective'. This social dimension meant that some people supported the event despite their wider concerns about park commercialisation: 'I have a problem with the growing corporatisation of public spaces generally and Victoria Park in particular. But private events like Lovebox for me are fine (and fun) because they host great music and bring a lot of people together to enjoy the park'.

The views outlined above highlight the issues facing large, well located parks in London which are increasingly used to stage large-scale music festivals. The event has now been moved to Gunnersbury Park in Hounslow, and the same debates have re-emerged there about the controversial transformation of this Park into a venue for a large-scale music festival. Similar events also take place in Hyde Park (City of Westminster), Trent Park (Enfield), Brockwell Park (Lambeth) and Finsbury Park (Haringey). These commercial incursions are justified as ways to generate much needed revenue for local authorities, and they are supported by a section of local residents who like the opportunities for entertainment they provide. This builds affinity for old-fashioned parks amongst young residents. However, these events are extremely divisive as they involve a trade-off between income generation and the accessibility/integrity of park spaces. Large-scale music festivals are one of the few types of event that can bring in sufficient amounts of money to help with the financial crisis affecting local authorities, but these events are also the most controversial and disruptive. This illustrates the unenviable dilemmas faced by local authorities tasked with maintaining London's parks.

Governance

The increased number of events staged in London's parks affects the ways that London's parks are used but it is also beginning to affect the ways they are governed. The potential to generate revenue from events – and the difficulties maximising and ring-fencing income within conventional local authority structures – mean that new organisations are being created. For example, in 2015 Wandsworth Council created a new company called Enable Leisure and Culture and awarded it a four-year contract to manage its parks and cultural services. The company is set up as a staff mutual – incentivising staff to increase revenue and cut costs – and it aims to become like other social enterprises that have grown by bidding for service contracts in other local authorities. Enable Leisure and Culture has a strong events focus, and one of the reasons it was established was to allow the Council's existing parks and events teams to work closely together within a new structure that allows the income earned from park events to be spent on parks. This new company illustrates how the new focus on events leads to shifts in the ways parks are governed and managed.

In other cases, rather than responsibility for all the parks in a Borough being outsourced to a separate company, more focused organisations have been established to manage individual parks. Potters Fields Park in Southwark is one such example. Before it was redeveloped into a more formal space this former bomb site was a wildlife park, run by a charity that managed temporary parks on undeveloped land. The transformation of loose space into a tightly landscaped park reflects the transformation in governance arrangements. Potters Fields Park is now managed by a dedicated Trust which generates its own revenue and ring-fences this money to be spent exclusively on the Park. The land is still publicly owned and is subject to Southwark Council's byelaws, but Potters Fields Park does not need any public funding for maintenance because of the income it generates. Over two thirds of this income is earned through events and the Trust employs two members of staff to manage park events that are staged on up to 56 days per year. The Park is located next to Tower Bridge and this creates demand from companies seeking to stage product launches and other commercially-oriented events there. This seems like an efficient way of funding a park, but the regularity with which these events are staged undermines the notion that this is open and public space. Ultimately, Potters Fields Park represents a new model of public space provision where parks are funded by allowing them to be privatised temporarily.

The new structures outlined above represent typical examples of the ways public service management is changing in the era of neoliberalism. These arrangements are criticised by many commentators who feel they undermine the democratic tradition of local government and encourage a culture where management is driven by financial motivations. The new arrangements have significant social justice implications. Some parks are more able to generate more commercial funding than others, and these parks tend to be those that are located in affluent areas. Therefore, the rise of commercial funding and associated governance structures are likely to exacerbate the inequitable access to quality park space that already exists in London (London Assembly 2017). Rather than detaching 'prosperous parks' from local authority control, as has happened in Southwark, it seems fairer to redistribute income earned by parks to those that are not in a position to generate large amounts of commercial revenue. This approach – adopted by Wandsworth Council – prevents the pernicious mode of park neoliberalisation criticised by Millington (2015) and other authors. However, even this model creates problems as there is a temptation to use high profile parks as revenue generating 'cash cows' to subsidise others. Perhaps the best example is Finsbury Park where critics have summarised Haringey Council's approach to park management as: 'We need the money, therefore we have to sacrifice Finsbury Park for the good of the rest.' (Interview with Chair of London Friends of GPN [in 2016]).

Regulation

One way of controlling the event takeover of London's parks is through effective regulation. Staging events in London parks is regulated through various planning and licensing requirements, giving local authorities the opportunity to ensure that park space is not overwhelmed or denigrated by inappropriate events. Many London parks are now licensed premises, but anyone seeking to stage an event in a public park still needs permission from the local authorities – with decisions guided by outdoor events strategies or dedicated park event policies. The latter are now produced by several London boroughs to control the number, size and nature of events that can be held in specific parks. For example, Enfield's new Parks Events Strategy 2017–2022 limits the number of events staged in each of the Borough's parks to eight in small parks and ten in larger ones. The timing of events is also controlled by these policies. Several of London's local authorities stipulate that major events (those catering for more than 5,000 people) cannot be staged in the school summer holidays and some policies indicate there must be a certain amount of time between major events (e.g. Hounslow). If event structures are to remain in place for more than 28 days, or if they are particularly extensive, then planning permission is also required. The obvious issue with all these regulatory mechanisms is that they are controlled by local authorities – but we know that these authorities are desperate to generate income to offset budget cuts. However impartial and scrupulous their planning and licensing procedures are, there is an inherent incentive to sanction lucrative events (Smith 2016).

There is also other legislation that is designed to help protect public open space from excessive commercial use. The Greater London Parks and Open Spaces Order 1967 applies to local authority owned parks and, whilst it permits the provision of 'amusement fairs and entertainments', it stipulates that spaces enclosed or set apart 'should not exceed in any open space one acre or one-tenth of the open-space, whichever is the greater'. This legislation seems to protect London's parks from excessively large events that take up an unreasonable amount of park space. How often 'amusement' can be staged is also regulated: the Order states this must be limited to 35 days and to a maximum of eight Sundays. Commercialisation is specifically regulated by the stipulation that 'the areas occupied by the paraphernalia of sales must not exceed one tenth of the area of the open space occupied by the function in question'. The 1967 Order anticipated the potential for conflict between event uses and everyday uses of London's parks – providing useful legislation to regulate competing demands for park space in the contemporary era.

Figure 10.4: Fences erected to stage Music Festivals in Hyde Park every summer (Photo: Andrew Smith).

However, several large events staged in London in recent years seem to contravene this legislation. Formula E events in 2015 and 2016 enclosed over 90 per cent of Battersea Park for four days and approximately 27 per cent of Finsbury Park is used for the Wireless Festival. This was the basis for the legal action undertaken by The Friends of Finsbury Park in 2016 when they applied for a Judicial Review of Haringey Council's decision to permit the Wireless Festival. The judge overseeing the case ruled that subsequent legislation (The Local Government Act 1972) meant the Council were entitled to stage this event. This decision was upheld in the Court of Appeal and so it would appear that London has lost the protection afforded to it by the Greater London Parks and Open Spaces Order 1967. A laissez faire regulatory landscape where local authorities can choose to do anything they like with parks they control means London threatens to overtake New York at the vanguard of park neoliberalisation. Whilst a series of large-scale music festivals have been sanctioned in London's parks, New York's Parks Department have adopted a much more cautious approach. In December 2016, three separate applications to stage music festivals in Corona Park – from AEG, Live Nation and the Madison Square Garden Company – were turned down. The reasons cited by the Commissioner of the Parks Department reflect the arguments made by event opponents in London:

Given the proposed duration of your three-day festival and the large amount of the Park that would be occupied for an extensive period of time, including the load in, loud out and the actual event, the Department has determined that the Park is not a viable venue for an event of this size and duration.

Conclusions

This chapter has identified how and why London's parks have become more intensively used as venues for commercial events. Urban parks are notoriously contested spaces, and the rise of commercial events adds to the reputation of parks as disputed territories. These conflicts are perhaps best understood as inevitable struggles between interests that value parks for their everyday use value, and those that seek to realise the exchange value of parks. The latter include event organisers, who use park venues to add value to their events, and local authorities seeking to generate revenue from public assets. Staging events has helped London's local authorities generate much needed income, but there is a danger that some Boroughs are now overly reliant on single events or over-exploited parks. Events are a relatively unreliable source of revenue given their high failure rate and due to the growing competition between park venues for events. There are other issues too. As with other forms of commercial income, it is not always clear how money earned from events is spent – and this under-mines the notion that events are justified because they help to pay for parks.

There is a danger of over-exaggerating the 'threat' posed by events, but the increase in the number and scale of events has important implications for Lon-don's parks. First, it affects the physical, symbolic and financial accessibility of much needed green space. Every time a ticketed event is staged the amount of genuinely public space available to use is diminished. These events com-municate the message that parks can be bought and fenced off, and these bar-riers erode the public feel and visual appeal of parks. Second, events – and the assembly/de-rig work needed to stage them – make London's parks more like the rest of the city, i.e. dominated by commercialism and construction work. Whilst this may please those stakeholders seeking to ensure Victorian-era parks remain relevant in the twenty-first century, it erodes London's reputation as a city punctured by green havens. And, third, whilst events are temporary, they can also have more permanent effects on the parks that host them. By normalis-ing commercial activity in parks, they provide precedents for further commer-cialisation and their increasing influence over new governance arrangements also represents a longer-term legacy. The rise of events also contributes to the broader commercialisation agenda as they are used as vehicles to enhance the visibility and image of parks. As parks are generating more income from com-mercial sources, there is an incentive to attract more tourists – and events are a good way of attracting the attention of these audiences. In this sense, events help to integrate parks into the wider visitor economy, transforming London parks from amenities into destinations.

London's parks should continue to stage well managed events, but the discussion in this chapter shows there is a need to protect parks from over-exploitation and over-commercialisation. Limiting the amount of park space and the number of days that major events are allowed to occupy seems justified; and these limits should incorporate days when events are being set and taken down. It is important to maximise the amount of public space still available during events and key facilities – playgrounds, sport facilities – should remain accessible. To avoid some of the conflicts seen in London during the summer of 2016, there needs to be more input from local user groups into decision-making about events/ event policies and better communication with residents about what events are happening and how they will affect parks. More transparency about how much money is generated by events and where this income goes would also help to better justify many of the large-scale events that are now staged in London's parks.

Staging major events in London's parks is a contested topic which divides opinion. A useful way of summarising the different arguments made to justify or resist this trend is through references to openings and closures. Opponents feel that events close down parks, disrupting use and restricting access – changes which they feel undermine the publicness of parks. Advocates argue the opposite – suggesting that events help to keep parks open – by generating much needed funds and by opening up traditional parks to new uses and users. Hence, event advocates argue that events make parks more public. One way to reconcile these contrasting views is to understand park events as agents of de- and re-territorialisation, in other words as interventions that both open up and close down public space. Music festivals, motor races and other large-scale events de-territorialise parks by challenging established meanings and identities, but they also re-territorialise parks as spaces of consumption for non-local users: tourists and visitors. This expansion and extension of commercial activity into London's green spaces is driven by the policies and rhetoric of neoliberal austerity – a context in which the traditional way of funding parks (through taxation and public finances) – is no longer deemed viable. In this sense, discourses of crisis are once again being used as a vehicle to push through changes to public park management.

References

Brenner, Neil and Nick Theodore. 2002. 'Cities and the Geographies of "Actually Existing Neoliberalism"'. *Antipode*, 34(3), 349–379.

Brent Council. 2016. *Report for Scrutiny on Civic Enterprise (Income Generation)*. Available at http://democracy.brent.gov.uk/documents/s45897/Income%20 Generation.pdf (accessed 7 December 2018).

Davidson, Mark. 2013. 'The Sustainable and Entrepreneurial Park? Contradictions and Persistent Antagonisms at Sydney's Olympic Park'. *Urban Geography*, 34(5), 657–676.

Event Lambeth. 2015. *A New Events Strategy 2015–2020*. Available at https://www.lambeth.gov.uk/sites/default/files/new-events-strategy-2015-to-2020.pdf (accessed 1 September 2017).

Glover, Troy. 2015. 'Animating Public Space'. In Sean Gammon and Sam Elkington, eds, *Landscapes of Leisure*. London: Palgrave Macmillan UK.

Haringey Borough Council. 2016. Freedom of Information Request Response. Available at https://www.whatdotheyknow.com/request/financial_implications_of_wireless festival (accessed 7 December 2018).

Heritage Lottery Fund. 2016. *State of UK Parks 2016: Research Report*, September.

Kohn, Margaret. 2004. *Brave New Neighborhoods: The Privatization of Public Space*. London: Routledge.

Krinsky, John and Maud Simonet. 2011. 'Safeguarding Private Value in Public Spaces: The Neoliberalization of Public Service Work in New York City's Parks'. *Social Justice*, 38 (1/2 (123–124)), 28–47.

Lewisham Council. 2016. *Income Generation Opportunities Review*. London: Lewisham.

London Assembly. 2017. *Park Life: Ensuring London's Green Spaces Remain a Hit with Londoners*. Available at https://www.london.gov.uk/about-us/london-assembly/london-assembly-publications/park-life-ensuring-green-spaces-remain-hit (accessed 7 December 2018).

Loughran, Kevin. 2014. 'Parks for Profit: The High Line, Growth Machines, and the Uneven Development of Urban Public Spaces'. *City and Community*, 13(1), 49–68.

Madden, David. 2010. 'Revisiting the End of Public Space: Assembling the Public in an Urban Park'. *City and Community*, 9(2), 187–207.

Millington, Nate. 2015. 'From Urban Scar to "Park in the Sky": Terrain Vague, Urban Design, and the Remaking of New York City's High Line Park'. *Environment and Planning A*, 47(11), 2324–2338.

Nesta. 2013. *Rethinking Parks: Exploring New Business Models for Parks in the 21st Century*. Available at https://www.nesta.org.uk/report/rethinking-parks-new-business-models-for-parks/ (accessed 7 December 2018).

Nesta. 2016. *Learning to Rethink Parks*. Available at https://www.nesta.org.uk/report/learning-to-rethink-parks/ (accessed 7 December 2018).

Perkins, Harold. 2009. 'Turning Feral Spaces into Trendy Places: A Coffee House in Every Park?'. *Environment and Planning A*, 41(11), 2615–2632.

Pine, Joseph and James Gilmore. 1999. *Experience Economy-Work is Theatre & Every Business a Stage*. Boston, MA: Harvard Business School Press.

Rosol, Marit. 2010. 'Public Participation in Post-Fordist Urban Green Space Governance: The Case of Community Gardens in Berlin'. *International Journal of Urban and Regional Research*, 34(3), 548–563.

Royal Parks, The. 2016. *Annual Report and Accounts 2015/6*. Available at: https://www.royalparks.org.uk/__data/assets/pdf_file/0005/64265/FINAL-Annual-Report-and-Accounts-2015-16.PDF (accessed 7 December 2018).

Smith, Andrew. 2014. '"Borrowing" Public Space to Stage Major Events: The Greenwich Park Controversy'. *Urban Studies*, 51(2), 247–263.

Smith, Andrew. 2016. *Events in the City: Using Public Spaces as Event Venues*. Abingdon: Routledge.

Smith, Harry, Dempsey, Nicola, and Mel Burton. 2014. 'Coordinating Place-Keeping: Towards More Sustainable Places'. In Nicola Dempsey, Harry Smith and Mel Burton, eds, *Place-Keeping: Open Space Management in Practice*. New York, NY: Routledge.

Wandsworth Borough News. 1991. 'Preserving the Patchwork'. *Wandsworth Borough News*. 8 November.

Zukin, Sharon. 1995. *The Cultures of Cities*. Oxford: Blackwell.

Conceptualising the Expansion of Destination London: Some Conclusions

Andrew Smith

> You are now
> In London, that great sea, whose ebb and flow
> At once is deaf and loud, and on the shore
> Vomits its wrecks, and still howls on for more
> Yet in its depth what treasures!
>
> (Shelley 1820)

Introduction

In the growing literature on city tourism a distinction is made between *tourism in cities* – that which incidentally occurs in urban environments – and *urban tourism*, where tourists are specifically interested in consuming urbanism (Ashworth and Page 2011). The preceding chapters suggest London tourism – particularly that which inhabits non-central areas – is dominated by urban tourism. This world city attracts people who want to be in London and experience the chaos and diversity of a postmodern metropolis (Gilbert and Henderson 2002). Whilst they might regard London as all consuming, these visitors are themselves consuming the city, a process that implies negative

How to cite this book chapter:
Smith, A. 2019. Conceptualising the Expansion of Destination London: Some Conclusions. In: Smith, A. and Graham, A. (eds.) *Destination London: The Expansion of the Visitor Economy.* Pp. 225–236. London: University of Westminster Press. DOI: https://doi.org/10.16997/book35.k. License: CC-BY NC-ND

consequences. As Hall et al. (2013) suggest, place consumption connotes degradation, destruction and displacement and there is evidence that in London increasing numbers of tourists (visiting a broader set of locations) have resulted in such consequences. Examples noted include the impacts of tourist rentals on the availability/affordability of housing (Chapter 3), the commodification and denigration of park environments (Chapter 10) and creeping touristification, which makes peripheral neighbourhoods less liveable (Chapter 2). In qualifying these effects, it is important to emphasise that tourism is merely one contributory factor. The problems of gentrification, commodification and homogenisation are caused by a combination of tourism working alongside other agents of change.

Most academic critiques of city tourism and its consequences tend to be quite negative, but we should not be too quick to label tourism in London as a problem. This is the city's second biggest 'industry' after financial services and one that supports thousands of jobs. Despite its negative connotations, tourism consumption is not necessarily a bad thing. As Hall et al. (2013) note, it may entail acts of assessment, appreciation, affection, assimilation, even enlightenment. Tourism in London, particularly 'new urban tourism', allows people to encounter difference, to share space with people from different backgrounds and to demystify 'the Other'. London is not immune to racism and religious intolerance, but the UK capital is generally recognised as an example of a super-diverse city where different people co-exist in relative harmony. The tourism slogan adopted by London and Partners, 'See the World Visit London', may be overselling the city's global credentials, but it contains an essential truth. People can understand the world better by spending time in London.

The other reason we should be wary of dismissing tourism as a problem is that its reputation as a parasitic activity is often exaggerated. Tourists are not merely consumers of city destinations, they are contributors: they not only bring income and investment; their very presence animates places (Pappalepore and Smith 2016). Rather than decrying the erosion of London's culture by outsiders, Newland (2008: 231) argues that 'tourists and immigrants continue to bring London to life'. This point is also made by Gilbert and Henderson (2002: 130) who feel that in London 'the tourist is an active and necessary part of the drama'.

In cases of overtourism, where visitor numbers have outgrown the capacity of destinations to host them, the obvious response has been to try to disperse tourism into more peripheral zones. This has been an ambition of policy makers in London for several decades. In recent years, non-central areas of London have certainly hosted more tourists. However, the increasing popularity of the periphery has supplemented, not substituted, rising demand for central districts. This is not tourism dispersal, it is tourism expansion. The ten preceding chapters of this book have helped to enhance our understanding of how this expansion has occurred and in this concluding chapter these ideas are drawn together into an overarching conceptualisation. Based on the work in this book, it would appear that tourism expands in a city destination *spatially, conceptually* and *temporally*. These three types of expansion are discussed further below.

Spatial Expansion

As Robert Maitland argues in Chapter 2, London's peripheral neighbourhoods are becoming popular destinations, a process driven by visitors' search for distinctiveness. Central London is regarded by some tourists as an environment 'staged' for tourists, rather than an authentic experience. This is emphasised by descriptions of central London as 'a fairy tale city of delights' that is regarded as a separate, tourism-oriented space even by people who live in London (Newland 2008). Therefore, increased tourist penetration of non-central districts (particularly those in in East London) can be interpreted as an attempt to access 'backstage' regions which better represent contemporary London. The irony is that this penetration inevitably paves the way for the touristification, commodification and homogenisation of these districts. Pioneering tourists then search further into the periphery to find distinctive and authentic neighbourhoods – perpetuating this cycle of change. This process has parallels with many other transformative cycles – including the way alternative cultures are co-opted by mainstream producers. The relevance of this analogy to the discussion here is highlighted by Gilbert and Henderson's work on London guidebooks (2002: 123):

> many of the places and activities identified in self-consciously alternative [London] guides published in the 1960s and 1970s were incorporated into guides aimed at a much broader audience in the 1980s and 1990s.

The authors suggest this did more than just change the way London was understood, it also 'fundamentally changed the nature of those places taken from the margins to the centre of the tourist experience' (Gilbert and Henderson 2002: 123).

In the past two decades places like Shoreditch, Spitalfields, Kings Cross and Borough have shifted from being niche destinations for alternative tourists, to mainstream sites that are part of general tourist itineraries. Places like Brixton, Bermondsey, and Peckham now seem to be undergoing the same transformations, with a whole series of peripheral neighbourhoods from Wood Green to Woolwich being touted as 'the new Shoreditch'. In this context, one can understand Robert Maitland's provocative prediction in Chapter 2 that this process will inevitably extend geographically beyond inner city peripheries to suburban locations. The current Mayor of London – Sadiq Khan – is from Tooting, a suburban district in South London. In a recent interview, he was asked what he would recommend tourists should do when they visit his local area and his answer revealed a lot about the tourism potential of London's suburbs:

> I think you'll find some of the best food in London, not just in the great curry houses on Tooting High Street but also in Tooting Market and Broadway Market. The Bingo Hall in Tooting has got a great organ and Frank Sinatra played there. Tooting Common, Wandsworth Common and Clapham Common are not far. But the people are the best thing about Tooting. (Khan cited in TimeOut 2018)

The combination of ethnic diversity, food markets, twentieth–century heritage and green space, alongside the capacity to interact with Londoners, represents the core appeal of many non-central districts. Following New York's example, London is increasingly exploiting diversity as a key tourism asset and suburban London has high levels of religious, ethnic and cultural diversity. There are also more traditional sightseeing opportunities. The most significant religious buildings built in London over the past fifty years are the temples and mosques located in the suburbs.

The potential of tourists to stay in peripheral destinations and experience London 'like a local' has been enhanced greatly by the rapid increases in the amount of accommodation available for tourists to rent. This phenomenon is explained by Clare Inkson in Chapter 3. Inkson describes this as an 'unplanned expansion' of accommodation provision – one akin to developing more traditional capacity without proper consideration of social impacts. The rise of Airbnb and other companies offering easy access to short term rentals has encouraged tourists to visit peripheral parts of London but it has come with considerable costs. These problems are caused by the failure of regulators to keep up with the new integration of the residential housing and tourist accommodation sectors.

The discussion above suggests that tourism territory expands like a frontier – creeping out in a radial fashion from the centre due to various demand trends, cost advantages and – to a lesser extent – tourism policy. This explains the main way that tourism territory is expanding in London. But there are other ways too. Chapters 4 and 5 by Anne Graham and Claire Humphreys discuss two examples of infrastructure that have provided the basis for new destinations in the urban periphery: airports and sports stadiums. There has been considerable investment in these types of projects in London over the past 20 years and whilst these facilities principally act as functional transport/leisure sites, they are now being redesigned to facilitate wider consumption and repurposed as catalysts for destination development.

Peripheral parts of London are attractive to some tourists because of their authenticity and everyday qualities, but more generic entertainment and consumption-oriented destinations are also emerging in London's periphery. Many of these are centred on airports or sports stadiums. For example, Wembley,

North Greenwich and Stratford are currently being redeveloped as significant destination zones with indoor and outdoor arenas acting as anchor projects. These places now provide cheaper and more convenient alternatives to the West End, serving densely populated residential zones nearby but also visitors coming from further afield – facilitated by advanced public transport infrastructure. These are UK versions of the suburban entertainment districts that are said to be emerging in the US. Spirou and Judd (2014) argue that suburbs and ex-urban locations are now investing heavily in projects that seek to emulate the success of city centre redevelopment schemes. These authors conclude that 'this functional refocusing of these cities may prove central to the next wave of urban change' (Spirou and Judd 2014: 46). In London, it seems unlikely that the dominance of the city centre will be challenged, but attempts to create new city centres (e.g. at Canary Wharf, Stratford and Croydon) may result in a revised spatial distribution of tourism that reflects this new urban structure.

The analysis contained within this book – particularly the work of Andrew Smith in Chapter 6 and Simon Curtis in Chapter 8 – also reveals a third type of spatial tourism expansion. Alongside the touristification of existing neighbourhoods (type 1) and the development of new purpose-built destination districts (type 2), we can see extensions which facilitate the consumption of central districts (type 3). Vertical extensions and opening up riverside spaces provide new ways of viewing the Cities of London and Westminster. For example, in Chapter 8 Simon Curtis notes how the developments along the South Bank have created new tourism areas for London, but their principal function according to Gilbert and Henderson (2002) is to direct the tourist gaze back towards the historic sights of the North Bank. In a similar manner, the new high rise structures that have been built in London over the past decade allow iconic buildings and spaces to be consumed more easily – providing a distanced perspective that is otherwise very difficult to achieve in a densely built-up city. New, dynamic experiences are offered atop new high-rise structures (e.g. slides, glass floors), but the principal attraction remains viewing London icons. This third type of spatial expansion is inherently connected to existing tourism centres: vertical and aquatic extensions reinforce the primacy of London's iconic districts rather than deconcentrating them.

Conceptual Expansion

The expansion of tourism territory is usually regarded as a physical and spatial phenomenon, but some of the chapters in the book highlight that tourism also expands by reaching into new spheres. There are two good examples of this in the book. In Chapter 3, Clare Inkson highlights that private homes are increasingly rented out to tourists. The appeal of this type of accommodation is partly based on the notion that tourists are able to penetrate the domestic realm – allowing them to experience London 'like a local' and 'feel at home' in the city. These ideas infuse much of Airbnb's advertising. Inkson rightly questions whether this actually happens, but the fact that tens of thousands of Londoners now offer their houses, flats or rooms for short term rent represents an extension of tourism – and capital - into the private sphere. The idea of the backstage again appears relevant here. In the contemporary era, tourists are not only interested in visiting 'real' neighbourhoods where Londoners live, they now want to stay in their houses. This is changing the dynamics of local neighbourhoods – creating new rhythms and rituals, but also disputes and displacements. London-based sociologist Lisa Mackenzie (2017) has tweeted about her experiences of this phenomenon, noting 'The Friday afternoon sound of wheeling suitcases all over East London as the airbnbs take over'.

The case of Unseen Tours discussed by Claudia Dolezal and Jayni Gudka in Chapter 7 represents another example of how tourists are being invited into secret/hidden worlds. Many of these tours traverse central places which are familiar to tourists: these are parts of London which are very much *on* the beaten track. Nevertheless, the insight and different perspective provided by guides allow familiar places to be seen in a different light, and overlooked features to be acknowledged. Uncovering secret or unseen places is a noted trend in city tourism. The enduring popularity of London's Open House event – when, for one weekend a year, people are able to access residences, offices and other private buildings – is part of the same trend. Whilst the penetration of tourists and tourism capital into the private sphere can be interpreted as the part of the ceaseless commodification of everything in the neoliberal city, the case of Unseen Tours suggest there can be more progressive outcomes.

Temporal Expansions

Popular destinations can manage high levels of demand by trying to disperse visitors over a larger area, but also by dispersing them into less busy time periods. The temporal expansion of tourism is also evident in London as noted by Humphreys in Chapter 5 and Eldridge and Pappalepore in Chapter 9. Demand is lower in winter months, so it makes sense to promote winter events, including sports events, but also Christmas-themed attractions and light installations. The Lumiere festival was created to provide a reason to visit London in January and the success of this event highlights the value of interventions that even out seasonal disparities. Eldridge and Pappalepore's discussion of Christmas lights also highlights a different type of temporal expansion – extending tourism into the night – something which has perhaps been neglected in existing analyses of urban tourism. Chapter 9 highlights the appeal of light-based attractions, but there are other ways tourists can enjoy London at night which extend beyond obvious night-time entertainment such as theatres, restaurants, bars and clubs. Many museums now offer regular late-night opening, with some (e.g. the Natural History Museum) offering sleepovers inspired by the popularity of the *Night at the Museum* book and film. In 2018 The Museum of London announced that their new building (due to open in 2023) will open 24 hours a day. Other famous attractions are also offering late night experiences, including London Zoo which offers Zoo Nights on Fridays during the summer months. The Royal Observatory in Greenwich offers regular stargazing events and there are a range of other tours (e.g. ghost tours, zombie experiences, wildlife spotting) which deliberately capitalise on nocturnal experiences and atmospheres.

Challenges

This book has focused on how London tourism is expanding, but it is perhaps useful to conclude with a more general assessment of the tourism related challenges facing the city. The city has undergone significant changes over the past two decades, with the new structures, attractions and governance introduced in 2000 providing the platform for the city's current popularity as a destination. Whether or not this success can be sustained in the face of competitive challenges and external factors is an important question. London is now accepted as one of the great world tourism cities, outperforming Paris and New York and only rivalled in terms of international arrivals by Hong Kong and Bangkok. But new competition is emerging in the Middle and Far East with Dubai, Abu Dhabi, Singapore, Kuala Lumpur and Singapore pursuing ambitious tourism targets. What will allow London to remain in the top tier of tourist cities, when faced with such well-endowed competition?

The key to London's success as a destination has been the way it has constantly added new layers of interest – often in new areas – to supplement established attractions and districts. Although there are many new projects planned, London (unlike its new rivals) does not have to expand tourism provision by building new districts. Instead, its existing, multi-layered urbanity can be further exploited. This is a significant competitive advantage over cities that are still being 'made'. Success will also depend on sustaining London's reputation as a world capital of culture – as a place where culture is produced rather than merely consumed. The 1960s were a key turning point for the city: this was when London became the place where fashions and tastes were defined (Gilbert and Henderson, 2002). This role has been retained ever since, and will need to continue if London is to provide new reasons for tourists to come – and to come back. Unfortunately, there are signs that London is becoming a victim of its own success. The price of property is now so high that many cultural pioneers are being forced to find cheaper places to live and work in other parts of the UK (e.g. Margate) or in rival cities on the continent (e.g. Berlin).

Economically, Destination London is a great success story, but the pressing challenge is to make tourism a more progressive force. Working conditions in the tourism and hospitality sectors remain a concern and plans for stricter controls on immigration may make it harder to fill positions. Ironically, whilst

Brexit may make it harder for London to find tourism staff, it is making London a more popular destination – at least in the short term. The falling value of the pound since the announcement of the referendum result in June 2016 means London has become relatively cheaper for most international tourists.

The ceaseless expansion of tourism makes it impossible to prevent tourism from influencing different districts. As Lim and Bouchon (2017) argue, tourism now permeates entire metropolitan areas and cannot be contained in dedicated bubbles. The city's ambition should not be to restrict tourism, but to better protect the integrity of urban districts from some of the negative effects of touristification. More regulation is required to try and safeguard community assets from real estate speculation, especially local pubs, small venues and independent shops. Ensuring London remains a lively and liveable city also makes it a more interesting place to visit. In Chapter 3 Clare Inkson discusses the need for tighter regulation of tourism rentals to protect the social fabric of local neighbourhoods. In London, there are now maximum limits on the number of days that home owners can rent out properties for short term lets (90 days), but much stricter limits apply in comparable cities like Amsterdam (60 days) and New York (30 days). This type of regulation is a key priority as the lack of affordable housing is perhaps the biggest challenge facing London: there are currently 243,000 people on the waiting list for social housing in the city. Aligning that figure with the vast number of properties now available for short term rent (over 60,000 are now listed on Airbnb), highlights that radical action needs to be taken to increase the availability and affordability of housing. Short term rentals are one small contributor to this problem, but a contributor nonetheless.

Instigating change will require organised and vocal opposition. Where developers have encountered resistance from those seeking to protect London from (over)commercialisation, there have been some significant victories. The famous Undercroft – the graffiti-strewn home of skateboarding in London – was threatened by the regeneration of London's South Bank district. But a campaign to keep the space succeeded and it is now legally protected as 'an Asset of Community Value'. In Greenwich, plans for a new cruise ship terminal were withdrawn after local campaigners challenged the project on the basis that it would lead to damaging levels of air pollution – another big issue currently facing London. These cases highlight that Londoners may need to follow the lead of their peers in Barcelona and Berlin – and actively campaign against key projects – if they want to restrict the over-expansion of the tourism sector.

In terms of managing visitor volumes there are already some measures in place to restrict overcrowding. For example, some London Underground stations near popular tourist sites restrict access at weekends (e. g. Camden Town and Covent Garden). There are also proposals to redesign some of London's most crowded areas to make them better able to cope with large numbers of people – including plans to pedestrianise Oxford Street. However, it seems likely that soft approaches – promotions, events and persuasion – will provide the main ways that London attempts to direct people away from overcrowded sites. Like in other cities, the official policy remains to disperse tourists into more peripheral districts and away from tourist hotspots. But this book suggests that, whilst it is possible to realise tourism growth in non-central areas, this doesn't necessarily mean tourism levels will decline in the centre.

Ultimately, to address many of the challenges facing Destination London, better tourism planning, and/or better integration of tourism considerations into urban planning are required. Just as tourism is neglected in the academic literature on cities and urban development, it is neglected in strategic planning processes. This situation seems to be getting worse. In the 1990s, 60 per cent of London's 33 Boroughs had a specific tourism policy or strategy but recent research suggests this figure has now dropped to 12 per cent, with only 4 of the 33 Boroughs producing a dedicated tourism plan (Maxim 2017). The integration of tourism into wider planning documents reveals an equally dismal picture. In 2000, over half of London boroughs had a dedicated chapter on tourism in their main planning document(s). That figure has now dropped to a measly 15 per cent (Maxim 2017). This means we now have a situation where tourism is growing rapidly in London, and expanding into more peripheral parts of the city, but fewer local authorities are doing anything to plan and manage it. Deficient planning has been caused by the massive cuts that Local Governments have had to endure over the past decade, with non-statutory functions like tourism jettisoned to shore up funding for front-line services. In the first decade of the twenty-first century, the London Development Agency (LDA) did undertake some useful tourism planning and development work, but since the LDA was abolished in 2012 London-wide planning and management has also been neglected.

In the absence of tourism planning by the public sector in most Boroughs (other than basic development control and licensing), this function has been delegated to Business Improvement Districts. These are partnerships of local interests who agree to pay a levy of between 1–2 per cent on the rateable value of their businesses. These monies are then ring-fenced to be spent on local initiatives such as improvements to the public realm, tourism promotions, event projects and safety initiatives. Fifteen BIDs were established by the LDA, but there are now over 50 operating in London. BIDS have introduced some welcome initiatives and helped to improve some areas (see Chapter 9), but the fact we are now relying on precarious, unaccountable partnerships to produce tourism plans for urban districts just demonstrates how far tourism planning in London has regressed. Whilst BIDs are well placed to assist business development, it is highly questionable whether they can advance social goals and progressive agendas. Therefore, revised governance arrangements – ones which not only give greater prominence to tourism, but allow tourism's social effects to be managed – are an urgent priority.

Mind the Gaps

The previous chapters in this book have highlighted a number of tourism-related issues facing London. However, the analysis here represents merely a sample of key challenges and trends and some important issues have been neglected in this volume. There are at least five notable omissions. First, by focusing on tourism spaces and places, the book has neglected the labour involved in servicing tourists and tourism. Work in the tourism and hospitality sectors is notoriously precarious, and the way working practices are changing in the gig economy requires detailed and careful analysis. A second key issue neglected by the book concerns ongoing restrictions that affect the accessibility of London. Tourism interests have long argued that it needs to be cheaper and easier for tourists to obtain visas to visit London – particularly the lucrative Chinese and Indian markets. The UK's proposed withdrawal from the European Union will also present new barriers that will also discourage international travel. A third key issue is the important influence of representations of London in literature, television and film. These are mentioned in Chapter 8 by Simon Curtis, but require further dedicated analysis. The worlds of Sherlock Holmes, Harry Potter and other fictional figures continue to motivate many tourists to visit London. The incongruous desire to see imaginary London in the flesh in perhaps best evidenced by the long queues to see the Platform 9 ¾ installation at Kings Cross railway station. Like many of the other examples discussed in this book, this type of tourism has the potential to push tourists and tourism into parts of London not normally regarded as visitor destinations. Leadenhall Market, Claremont Square and Stony Street are all non-central locations that have become visitor attractions thanks to the *Harry Potter* films.

A fourth issue not covered adequately in this book is the role of London as a business tourism hub and the wider relationship between London's business functions and its tourism sector. One notable development in this sphere is the attempt to reposition the City of London – the historic square mile where most financial institutions are based – as a visitor destination. This mission has inspired a project to develop a 'Culture Mile' between Moorgate and Farringdon. This is a joint initiative between The City of London Corporation and the main cultural institutions located within their territorial boundaries: The Barbican, The London Symphony Orchestra, the Museum of London and the Guildhall School of Music and Drama. In the past the dominant role of the City of London as a place of *work* meant it was always relatively quiet at weekends, but this is changing as this area is reimagined as a visitor destination. This district highlights the way urban areas are now being redesigned to service a mixture of office workers, citizens at leisure and tourists. Similar projects are being introduced at London's other financial services hub: Canary Wharf. Fifth and finally, this book has not properly addressed the changes to London's high streets caused by the rise of online shopping and ongoing economic malaise. Many of London's high streets, particularly those in peripheral districts, are struggling to maintain their traditional functions as places to shop. Given the significant role of the high street in London's peripheral districts, this trend may slow the emergence of many of these non-central districts as visitor destinations.

A Research Agenda

This book aimed to address the relative absence of work on London tourism noted by Maxim (2017) and other authors. However, further work is needed to develop some of the key issues raised. In particular, it is important to examine whether the types of tourism growth London is currently experiencing, and the ways tourism is expanding in the UK capital (spatially, conceptually and temporally), are relevant to other world tourism cities and other cities in general. The relationship between city tourism and urban change still needs to be better understood. This is a reciprocal relationship and needs to be analysed as such: city tourism contributes to urban change, but urban change also influences tourism. Understanding how tourism intersects with wider urban processes – of gentrification, globalisation and commodification – is crucial to understanding the role tourism plays in contemporary cities – particularly global cities like London. We also need to better understand how tourism relates to other urban activities, e.g. commuting and consumption – and how citizens experience their own cities as tourists.

The expansion of the visitor economy in London is contested, but more research is needed to understand why protests against this rapid growth seem to be much less significant in London compared to other world tourism cities like Berlin or Barcelona. The extreme volume of tourists in the UK capital means London provides an ideal laboratory to undertake much needed research on measures to reduce overcrowding. Future research on this theme will no doubt make use of the new ways of measuring, modelling and mapping crowds, including those based on mobile phone applications. Tourism is usually defined as an overnight stay by a non-resident and this book has also highlighted the need for more research on the increasingly varied ways that London and other cities are consumed at night. Finally, in an era of neoliberal austerity, when governments are constantly seeking new revenue streams to offset reductions in public funding, there is a need for more research on the potential for tourist taxes and other charges. As Pine and Gilmore (1999) point out in their famed account of the 'experience economy', the capitalist economy tends to expand by charging a fee for things that were once free. In this context, it seems likely that future London tourists will be asked to contribute to the upkeep of parks, museums and other public services.

References

Ashworth, Gregory and Stephen Page. 2011. 'Urban Tourism Research: Recent Progress and Current Paradoxes'. *Tourism Management*, 32(1), 1–15.

Gilbert, David and Fiona Henderson. 2002. 'London and the Tourist Imagination'. In Gilbert, Pamela, ed, *Imagined Londons*. New York: SUNY Press 121–136.

Hall, Michael et al. 2013. 'Vanishing Peripheries: Does Tourism Consume Places?' *Tourism Recreation Research*, 38(1), 71–92.

Lim, Sonia and Frederic Bouchon. 2017. 'Blending in for a Life Less Ordinary? Off the Beaten Track Tourism Experiences in the Global City'. *Geoforum*, 86, 13–15.

Mackenzie, Lisa. Tweet on 22 Sept 2017 (available at https://twitter.com/redrumlisa/status/911244351581429766).

Maxim, Cristina. 2017. 'Challenges Faced by World Tourism Cities – London's perspective'. *Current Issues in Tourism*, 22(9), 1–20.

Newland, Paul. 2008. *The Cultural Construction of London's East End: Urban Iconography, Modernity and the Spatialisation of Englishness.* New York: Rodopi.

Pappalepore, Ilaria and Andrew Smith. 2016. 'The Co-Creation of Urban Tourism Experiences', 87–100. In Antonio Russo and Greg Richards, eds, 2016 *Reinventing the Local in Tourism*. Bristol: Channel View.

Pine, Joseph and James Gilmore. 1999. *The Experience Economy. Work is Theater and Every Business a Stage.* Harvard: Harvard Business Review Press.

Shelley, Percy. 1820. Letter to Maria Gisborne.

Spirou, Costas and Dennis Judd. 2014. 'The Changing Geography of Urban Tourism: Will The Center Hold?' *disP–The Planning Review*, 50(2), 38–47.

TimeOut. 2018. Ask Sadiq. *TimeOut London*, 16 December.

The Editors and Contributors

Andrew Smith is a Reader in Tourism and Events at the University of Westminster, where he co-leads the Tourism and Events Research Group with Anne Graham. Andrew read Geography at the University of Cambridge before completing a PhD on city reimaging funded by both Universities in Sheffield. Since joining the University of Westminster in 2004 he has written extensively about urban tourism and supervised several PhD students working in this field. His research focuses on UK cities, but he has also published well-known papers on Barcelona, Oslo and Valletta. Andrew combines his work on urban tourism with a strong interest in city events and he has authored two books on this theme: *Events and Urban Regeneration* (Routledge, 2012) and *Events in the City* (Routledge, 2016).

Anne Graham is Professor of Air Transport and Tourism Management at the University of Westminster, where she co-leads the Tourism and Events Research Group with Andrew Smith. Before joining the University, Anne worked in air transport consultancy. She has two key research interests. The first is in airport management, economics and regulation. Her other research area is the analysis of tourism and aviation demand. She has published widely and her recent books include: *Air Transport: A Tourism Perspective* (2019); *The Routledge Companion to Air Transport Management* (2018); *Managing Airports: An International Perspective* (2018); *Airport Finance and Investment in the Global Economy* (2017); *Aviation Economics* (2016); *Airport Marketing* (2013); and *Aviation and Tourism: Implications for Leisure Travel* (2008).

Simon Curtis† was a Senior Lecturer in Tourism Management at the University of Westminster, 2011-2019. He was a specialist in destination management and tourism development with over 25 years of experience in the UK tourism industry before he became an academic. Simon began his career with the English Tourist Board specialising in product development and area regeneration initiatives. He then worked in management consultancy with Touche Ross and followed this with 15 years in senior positions in local government at Kent County Council and Medway Council. In Medway, he was Head of the Tourism and Heritage service which involved managing a department overseeing destination marketing, major events, the operation of Rochester Castle and the local museum and archive service.

Claudia Dolezal is a Lecturer in Tourism at the University of Westminster and is on the Editorial Board of the Austrian Journal of South-East Asian Studies (ASEAS). Her background is in tourism, international development and social anthropology, with a geographical focus on the region of Southeast Asia. Claudia's research interests focus on tourism and social inequalities, tourism for development and community-based tourism. She is specifically interested in the ways marginalised groups use tourism for empowerment, as well as the multi-faceted power dynamics that shape the tourism encounter in tourism settings of both the developed and developing world.

Adam Eldridge is a Senior Lecturer in Sociology in the School of Social Sciences at the University of Westminster. He has a PhD from the Department of Gender and Cultural Studies at the University of Sydney. His research examines the night as a context for new forms of sociality, leisure and urban development. His books include *Planning the Night-time City* (Routledge, 2010) and *Exploring Nightlife: Space, Society and Governance* (Rowman and Littlefield International, 2018).

Jayni Gudka leads Unseen Tours' Responsible Tourism and Communications work, as one of the organisation's (volunteer) directors. Unseen Tours is an award-winning social enterprise which offers unique London walking tours led by homeless, formerly homeless, and vulnerably housed tour guides. In her day job, Jayni develops community-based tourism micro-enterprises in Asia and West Africa. In her spare time, Jayni is also setting up the Travel CAST (Changing attitudes and stereotypes through tourism) Network – giving slum, homeless, refugee and other marginalised groups the opportunity to benefit from and enrich local travel experiences around the world through responsible tourism.

Claire Humphreys is a Visiting Lecturer at the University of Westminster. From 2011 to 2018 she was the Course Leader for the University of Westminster's highly regarded MA Programmes in Tourism Management and Events and Conference Management. She has experience working in and researching different aspects of the tourism industry – experiences which have informed the academic textbooks and journal articles she has written. Her PhD explored the influence of consumer decision making in sport tourism, which has shaped her recent work on stadium tours.

Clare Inkson is a Senior Lecturer at the University of Westminster where she teaches tourism at undergraduate and postgraduate levels. She has extensive professional experience in tourism with inbound and outbound tour operators, wholesalers, and agencies in operational and marketing roles. Her research interests include tourism distribution channels, destination development and marketing, and the tourism sharing economy. She is co-author of *Tourism Management – An Introduction* (Sage, 2018).

Robert Maitland is Emeritus Professor of City Tourism at the University of Westminster, where he founded the Centre for Tourism Research. An urban economist, his work focuses on how tourism shapes cities and cities shape tourism. His particular interests are the tourist experience of cities, in particular world cities, national capitals and heritage cities, with an emphasis on tourism and everyday life. He has advised government and industry agencies on these topics, and his books include *World Tourism Cities*; *City Tourism: National Capital Perspectives*; *Social Tourism: Perspectives and Potential* and *Tourism in National Capitals and Global Change*.

Ilaria Pappalepore is a Senior Lecturer in Events and Tourism at the University of Westminster. Her research interests include urban tourism, cultural events, creative industries and urban regeneration. Ilaria has worked on a variety of research projects, including a large-scale study on youth leisure funded by the Joseph Rowntree Foundation and an IOC-funded project on the impact of London 2012 on small creative organizations in East London. Prior to joining the University of Westminster, Ilaria worked with the Lille Development Agency, France, and later as Course Leader and Senior Lecturer in Tourism at Anglia Ruskin University in Cambridge.

List of Figures with Credits

List of Tables

Z

Printed and bound by CPI Group (UK) Ltd, Croydon, CR0 4YY

07/07/2026

14916228-0002